Advances in Experimental Medicine and Biology

Volume 1100

Advances in Experimental Medicine and Biology presents multidisciplinary and dynamic findings in the broad fields of experimental medicine and biology. The wide variety in topics it presents offers readers multiple perspectives on a variety of disciplines including neuroscience, microbiology, immunology, biochemistry, biomedical engineering and cancer research. *Advances in Experimental Medicine and Biology* has been publishing exceptional works in the field for over 30 years and is indexed in Medline, Scopus, EMBASE, BIOSIS, Biological Abstracts, CSA, Biological Sciences and Living Resources (ASFA-1), and Biological Sciences. The series also provides scientists with up to date information on emerging topics and techniques. 2017 Impact Factor: 1.760

More information about this series at http://www.springer.com/series/5584

Julio A. Aguirre-Ghiso
Editor

Biological Mechanisms of Minimal Residual Disease and Systemic Cancer

Springer

Editor
Julio A. Aguirre-Ghiso
Division of Hematology and Oncology,
Department of Medicine, Department
of Otolaryngology, Department of
Oncological Sciences, Tisch Cancer
Institute, Black Family Stem Cell
Institute, Mount Sinai School
of Medicine
Icahn School of Medicine at Mount Sinai
New York, NY, USA

ISSN 0065-2598 ISSN 2214-8019 (electronic)
Advances in Experimental Medicine and Biology
ISBN 978-3-319-97745-4 ISBN 978-3-319-97746-1 (eBook)
https://doi.org/10.1007/978-3-319-97746-1

Library of Congress Control Number: 2018958003

This Springer imprint is published by the registered company Springer Nature Switzerland AG
The registered company address is: Gewerbestrasse 11, 6330 Cham, Switzerland

Preface

Understanding and Targeting Minimal Residual Disease: A Path to Change Cancer Patient Outcomes

Technologies of early detection continue to improve, and therapies are now more than ever better tailored to treat specific cancers and their subtypes. With these improvements the survival curves for different cancers are showing rightward shift toward later relapse. In some cancers such as ER+ breast and prostate cancer, late relapses can span more than a decade and in some studies in ER+ breast cancer more than two decades. In hematological malignancies, late relapse spanning > 10 years is also documented. In diseases like melanoma, ultra-late recurrence is not uncommon, and even in HNSCC and renal cell carcinoma, these late recurrences are also observed. However, despite this progress, large fractions of patients bearing hematological and solid tissue malignancies relapse with refractory and lethal disease. This is due to the ability of residual cancer cells that do not proliferate to persist undetected by the immune system and unscathed by therapies.

Work by several labs in the last 20 years have shed light into the mechanisms by which cancer cells that otherwise should be in perpetual growth stop to enter a dormant state that then is interrupted to fuel relapses. The data has led to the notion that microenvironmental and epigenetic mechanisms may be responsible for maintaining residual cancer cells in a dormant state. These include control by the immune system or by niches that normally keep differentiated or adult quiescent stem cells in a growth arrest.

In patients, minimal residual disease (MRD) is monitored using different technologies. Hematological malignancies studies have pioneered the monitoring of residual cancer cells by specific markers combined with the detection of genetic alterations (commonly the driver mutated genes) that provide a molecular readout of the status of MRD. Other cancers such as prostate cancer rely on soluble markers in circulation such as PSA but do not directly interrogate the status of the residual cancer cells and their heterogeneity. This is in fact a major shortcoming in monitoring and understanding residual cancer cell biology in solid cancers. The detection of circulating tumor cells (CTCs, in blood) or disseminated tumor cells in some target organs (DTCs, in lymph nodes or bone marrow) has changed how we monitor the progression of disease. Still major efforts are needed to better characterize this residual disease to make clinical decisions and tailor therapies in order to prevent recurrences.

In this book nine chapters delve deep into the basic mechanisms of residual disease dormancy and recurrence in solid cancer and in hematological malignancies. These chapters have also dedicated effort to explain and analyze the methods and technologies being developed to monitor MRD in different cancers.

Pisarsky et al. explore the role of vascular biology in controlling metastatic progression and how anti-angiogenic therapies may in fact alter vascular homeostasis and awaken dormant residual cancer in various cancers, with an emphasis in breast cancer. Carlini et al. explore the basic microenvironmental and epigenetic mechanisms driving residual cancer cell dormancy with a focus on developmental mechanisms and therapy effects on dormancy programs. Cackowski et al. focus on MRD in prostate cancer and how bone marrow niches control DTC dormancy while also exploring the clinical perspectives of CTC detection. Lin et al. provide insight into MRD in both melanoma and breast cancer and how the use of CTCs cell-free DNA and cell-free microRNA detection are being adapted for monitoring residual disease in these cancers. With a focus in solid cancers as well Sproll et al. provide a detailed analysis of routes of dissemination that create MRD in HNSCC and esophageal cancer with the clinical perspective and mechanistic advances that are helping understand how to detect CTCs and DTCs and better manage recurrences in this cancer type.

Four chapters provide a comprehensive and insightful analysis of MRD in hematological malignancies covering acute myeloid leukemia (AML) (Gomez et al.), childhood acute lymphoblastic leukemia (ALL) (Jeremias et al.), chronic lymphocytic leukemia (Bhatia et al.), and multiple myeloma (Berger et al.). These chapters provide an in-depth look into the monitoring of MRD in the clinic. These chapters are revealing how advanced the detection of MRD is in hematological malignancies, compared to solid cancers. They also explain how this technology and consensus development in methods has benefited from a longer research experience and unified monitoring procedures that can be shared and compared across centers in the world allowing for consistent optimization and data sharing. These chapters also offer input into the biological basis of MRD in these hematological malignancies. They explore the genetic basis of these hematological malignancies and how they allow for detection of remnant clones and the role of the microenvironment and also stem cell programs in the control of MRD persistence in an inactive state followed by reactivation.

By bringing together experience in solid cancer, basic mechanism analysis of MRD, and dormancy and hematological malignancies studies, this book fills a niche for cross comparing how MRD is studied and understood in these cancers. The book not only provides biological insight but importantly a strong clinical perspective that must always guide the exploration of the biological mechanisms. Overall, the chapters in this book offer an insightful view of the current status of MRD and the mechanisms of dormancy and progression during the residual stage of the disease. Understanding this biology is key to target MRD and provide therapeutic opportunities to prevent recurrences. This is without doubt one of the biggest challenges in modern oncology that will provide improved outcomes for patients.

New York, NY, USA Julio A. Aguirre-Ghiso

Acknowledgments

I would like to thank the authors for their exceptional chapter contributions. Without these we could have not accomplished this project that brings needed new views and ideas to the minimal residual cancer field.

Contents

Epigenetic and Pluripotency Aspects of Disseminated Cancer Cells During Minimal Residual Disease

1

Maria Jose Carlini, Nitisha Shrivastava, and Maria Soledad Sosa

Abstract

Our understanding of the minimal residual disease (MRD) in solid cancers indicates that it can persist in the system for years or even decades. We now know that the persistence of MRD might depend on the dormancy of the disseminated cancer cells (DCCs). Once DCCs exit dormancy, they become metastatic and the survival rates of the patients inevitably decrease. Thus, innovative treatments are required to extend the asymptomatic phase of MRD after the initial therapeutic intervention. With the latest advances in cancer research, there is a greater need to explore and understand the biology, timing of dissemination, and origin of DCCs during tumor progression. These important aspects of DCCs impact the selection, design, administration, and timing of effective therapies. Herein, we summarize the current understanding of MRD biology in solid tumors, with a focus on epigenetics and pluripotency, presenting an overall view of the direction the field is taking to reach the goal of reducing cancer-related mortalities that result from metastasis.

Keywords

Quiescence · Dormancy · Pluripotency · Epigenetics · Minimal residual disease · Early dissemination · Metastasis · Disseminated cancer cells

Abbreviations

ADT	androgen depletion therapy
AR	androgen receptor
ATG	autophagy-related gene
atRA	all trans retinoic acid
BET	bromodomain and extraterminal domain
BM	bone marrow
BMP	bone morphogenetic proteins
BP	bisphosphonate
CAF	cancer associated fibroblasts
CCC	circulating cancer cells
CNV	copy number variations
CSC	cancer stem cells
CUP	cancer of unknown primary origin
DCC	disseminated cancer cells
DNMT1	DNA methyltransferase 1
E2	estradiol
ECM	extracellular matrix
EGFR	epithelial growth factor receptor
EMT	epithelial-to-mesenchymal transition

M. J. Carlini · N. Shrivastava · M. S. Sosa (✉)
Department of Pharmacological Sciences, Icahn School of Medicine at Mount Sinai, New York, NY, USA
e-mail: soledad.sosa@mssm.edu

J. A. Aguirre-Ghiso (ed.), *Biological Mechanisms of Minimal Residual Disease and Systemic Cancer*, Advances in Experimental Medicine and Biology 1100,
https://doi.org/10.1007/978-3-319-97746-1_1

ER	estrogen receptor
ESC	embryonic stem cell
GHB	4-hydroxybutyrate
HDAC	histone deacetylases
HER2	human epidermal growth factor receptor 2
HF	hair follicle
HNSCC	head and neck squamous cell carcinoma
HPSCC	hypopharyngeal squamous cell carcinoma
iPSC	induced pluripotent stem cells
LA	leukapheresis
LOH	loss of heterozygosity
MALAT	metastasis associated lung adenocarcinoma transcript 1
MMPs	matrix metalloproteinases
MRD	minimal residual disease
OSKM	Oct3/4, Sox2, Klf4, and Myc transcription factors
PGR	progesterone receptor
RARβ	retinoic acid receptor β
SAM	S-adenosylmethionine
SOX9	SRY [sex determining region Y]- box 9
SPARC	secreted protein acidic and rich in cysteine
SSADH	succinic semialdehyde dehydrogenase
TET	Ten-eleven Translocations
TF	transcription factors
TGF-β	transforming growth factor-β
TSS	transcription start sites
αKG	α-ketoglutarate

1.1 Introduction

Despite our knowledge of the basic biology of cancer and advances in diagnosis and treatment that have resulted in the extension of disease-free survival periods, the mortality rate for neoplasms has only modestly declined in the last four decades [1]. This may be because research has mainly focused on primary tumors instead of on metastasis, which is the predominant cause of mortality in cancer. Disseminated cancer cells (DCCs) are considered the precursors of metastasis and manifest no symptoms of disease for prolonged periods of time [2–5]. DCCs are cancer cells that disseminate from the primary sites, enter the circulation, and become embedded in the secondary organs. At these distant sites, DCCs undergo a non-proliferative dormancy program from which they eventually exit and reactivate to form metastases. Data from the last 10 years suggest that the biology of a single DCC, or of clusters consisting of less than 20 cells, differs from that of micro-metastases and metastases [3–6]. However, several questions still need to be answered in order to design treatments that might prevent metastases, such as: How is DCC biology different from that of primary tumors? Where do DCCs originate from? How can we avoid DCCs from becoming metastatic?

1.1.1 Minimal Residual Disease and Dormancy

In solid cancers, after removing the primary tumor, cancer cells can be eliminated to the extent of being undetectable with traditional tests (radiographic analysis and serological tests). Nevertheless, a small number of cancer cells at the primary site or disseminated to other organs can remain after treatments and these constitute a minimal residual disease (MRD). The widespread presence of DCCs has been shown in animal models [6, 7] and autopsied patients [8]. However, only a few sites can be safely sampled from living cancer patients to detect DCCs, such as bone marrow (BM) aspirates [9–11] or the lymph nodes [12–14].

At the single cell level, dormancy of DCCs is defined as cells that enter quiescence (G0-G1 phase of cell cycle) and can survive stress conditions. At the population level, dormancy implies the presence of balanced tumor cell proliferation and death ratios, because of either the insufficient availability of oxygen and nutrients (angiogenic dormancy) or the immune system control, which restrains the tumor mass size (immune system-regulated dormancy) [4, 15]. Dormancy is an actively maintained reversible state and it is therefore desirable to eradicate dormant DCCs or perpetuate their dormancy phase, since they might act as reservoirs for future relapses. In nature, it is known that embryonic stem cells (ESCs) from certain mammals can be induced into deep quiescence, which can be reactivated after months with the capacity to develop into all the functional tissues of an adult animal [16], a process known as diapause. Thus, it is not inconceivable that cancer cells could spontaneously mimic the same properties as ESCs to control their fate or that we can manipulate them into quiescence. Before tumor cells reach a secondary organ and become DCCs, they travel through the circulatory system as circulating cancer cells (CCCs). CCCs can also be detected in patients in remission, suggesting that DCCs from bone marrow compartments or other reservoir organs enter circulation even during MRD [17, 18].

Several clinical cases have provided evidence for the dormancy of DCCs [19, 20]. For instance, in 2010, individuals with over 10 years of MRD became eligible organ donors because of the high demand for organs at that time. As a result, recipients of these organs developed metastases from the previous diseases of the donors [19, 20]. This finding proved that MRD is a period in which dormant DCCs exist and can lead to future relapses when the right signals are present. Even though the presence of DCCs in the bone marrow of cancer patients is known to constitute a risk for relapse after surgery [9, 21–23], what controls the timing for this reactivation and its regulatory factors still remain unknown. From a clinical perspective, it will be important to identify molecular markers on DCCs and CCCs that identify patients with short- or long-term MRD and to select suitable dormancy-inducing therapies [24] to prevent reactivation. For instance, the BM DCCs of clinically dormant prostate cancer patients (with 8 to 18 years in remission) showed an enriched dormancy-related signature identified through immunofluorescence and single cell gene expression analysis [10]. This signature included *NR2F1* and *BMP7* genes, which will be described later as important regulators of dormancy. Whether this transcriptomic signature can be used to monitor disease progression requires further examination.

1.1.2 Early Dissemination Events: New Sources of Dormant DCCs

The latest research on early dissemination events has increased our understanding of not only the biology of DCCs but also their complexity. For years, the scientific belief was that tumor dissemination occurred at the invasive front of primary tumors. However, a large amount of evidence now suggests that dissemination takes place before the detection of invasive tumors (e.g., ductal carcinoma in situ, DCIS [25]) and at locations not related to the invasive front identified by pathologists [26]. Cancers of unknown primary (CUP) are another example of the existence of early dissemination or dissemination that does not require a detectable invasive tumor [27]. Genomic analysis of bone marrow DCCs from melanoma and breast cancer have proven the existence of early dissemination events [11, 25, 27–29]. For instance, the genomic analysis of DCCs derived from breast cancer patients [11, 28] identified DCCs with early aberrations and less progressed genomic changes than in the invasive primary tumors. Recent studies have also revealed some mechanistic insights into how early dissemination occurs [30–34]. An epithelial-mesenchymal transition (EMT)-like program and resistance to apoptosis was found to favor dissemination at the early stages of tumor progression (i.e., mammary intra-epithelial neoplasia in mouse models that resembles ductal carcinoma *in situ* in humans) [31]. It has also been shown that progesterone-induced signaling favored early dissemination [32]. Interestingly, increased levels of human epidermal growth factor receptor 2 (HER2) mitigated, through microRNAs and activated proliferation signals, the progesterone-induced dissemination [32]. Moreover, it was recently found that macrophages orchestrated the early dissemination process by inducing EMT in early breast cancer cells [30]. These programs provided mammary epithelial cells with the ability to migrate, enter the circulation, and colonize target organs. Early DCCs remain dormant for extended periods of time and have the capacity to eventually initiate metastasis [31]. Interestingly, the dormancy phase of early DCCs seems to be independent of p38 signaling, which argues for a different dormancy biology than that observed in late DCCs [31]. Furthermore, Hosseini et. al. showed that approximately 80% of metastases found in a HER2-driven mouse breast cancer model were derived from early cancer cells [32]. In addition, clinical evidence supports the idea that early DCCs derived from DCIS lesions are responsible for breast cancer mortality [35]. Thus, it may be possible that early DCCs persist during MRD and contribute to late relapses. These findings raise new questions: Do early DCCs enter a more resistant and long-lasting dormancy phase than do late DCCs? Could we correlate the presence of early aberrations in DCCs with dormancy susceptibility? Do DCCs derived from different stages of cancer depend on each other to allow for outgrowth or to promote resistance to drugs? Will eliminating a certain population of DCCs in patients avoid later recurrence? The latter has been partially addressed by the work of Linde et al. demonstrating that the elimination of early DCCs reduced later metastases in an animal model [30]. In this study, the authors showed that the elimination of macrophages only during early stages of tumor progression (before detectable tumor masses are formed) decreased the number of solitary DCCs in lungs and lung metastases in a HER2-driven mouse model [30]. These results argue that the elimination of DCCs, in this case early DCCs, affects metastasis development later in cancer progression.

In this chapter, we discuss studies that have described the biology of DCCs. In the first part, we debate the epigenetic programs regulating the dormancy of DCCs. In the second part, we comment on the pluripotent features of DCCs. Overall, the literature discussed here investigates DCC and CCC origins and their biology in an effort to guide development of efficient systemic therapies.

1.2 Epigenetic Mechanisms that May Control DCC Dormancy During MRD

Collective evidence suggests that the reversibility of the dormancy program in DCCs may be controlled by epigenetic regulations [6, 36–38] (Fig. 1.1). Changes in the DNA methylation patterns, histone modifications, histone variants, long non-coding RNA, and complexes involved in chromatin accessibility are able to shape chromatin structure and control gene expression. This implies that certain signals, for instance those coming from target organs, or intratumoral signals, might trigger changes in the epigenome of DCCs together with previously described pathways [4], which leads to dormancy. These epigenetic changes can be reverted by other signals to allow DCCs to exit dormancy.

1.2.1 Extracellular Matrix, Stem Cell Niche, and Aging Shape the DCC Epigenome

The DCC niche is structurally supported by the extracellular matrix (ECM). Beyond its supportive role, the ECM can directly initiate signaling events by acting as a precursor of active fragment derivatives [39]. In this context, changes in the ECM and tissue architecture can also regulate gene expression through chromatin remodeling [40–42], thus playing an important role in determining DCC fate. It was shown that architecturally intact normal fibroblast-derived ECM induced the translocation of the histone demethylase JMJD1a from the nucleus to the cytoplasm, followed by a downregulation of JMJD1a and its proliferation-inducing genes targets, such as YAP/TAZ [43]. When DCCs encounter a new environment, they may undergo chromatin remodeling induced by normal stroma combined with intrinsic DCC signaling that initiates dormancy.

Interestingly, factors secreted from primary tumors may shape the ECM at target organs by, for instance, promoting the localized deposition of fibronectin [44]. Fibronectin, along with

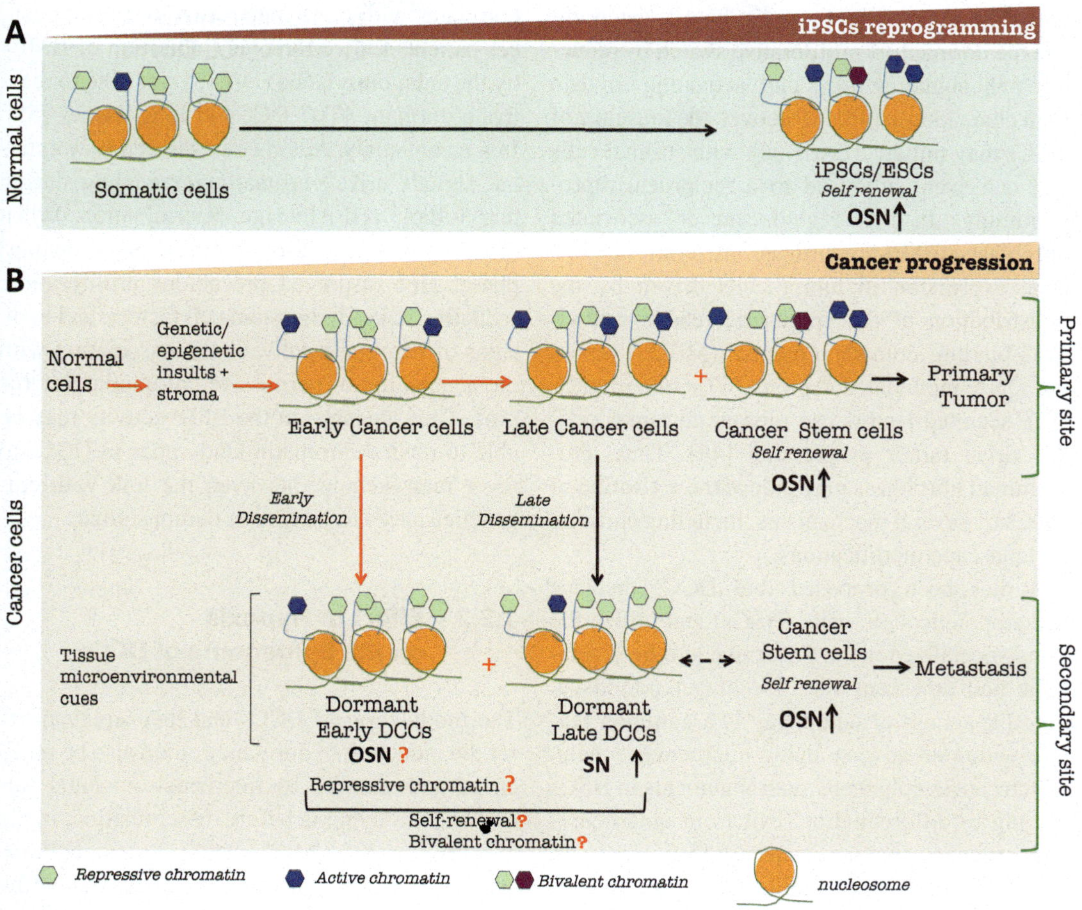

Fig. 1.1 Comparison of global chromatin landscape during induced pluripotent stem cells (iPSCs) reprogramming (Panel A) and cancer progression (Panel B). (a) Chromatin state in somatic cells and iPSCs. Somatic adult cells present a compacted chromatin with repressive histone marks (green hexagons), which acquire a more accessible open configuration with active marks (blue hexagons) and bivalent chromatin (purple and green hexagons) during iPSCs reprogramming [132]. Fully reprogrammed iPSCs are epigenetically identical to embryonic stem cells (ESCs) [132]. The generation of iPSCs requires OCT4, SOX2, c-MYC and KLF4 [95] expression. Forced *NANOG* overexpression leads to fully reprogramed iPSCs [133] (*OSN*: *OCT4, SOX2,* and *NANOG*). **(b) Epigenetics during cancer progression.** Early cancer cells may present a chromatin structure more similar to that in normal cells (**Panel A**) than in late cancer cells or cancer stem cells (CSCs). For simplicity, in this scenario we consider that early cancer cells originate from normal cells, including stem or progenitor cells, but the exact origin needs to be identified. CSCs share bivalent chromatin [134, 135], express pluripotency-associated transcription factors like OCT4, SOX2, and NANOG [136] (OSN), and have the capacity to form tumors. Parallels have been drawn between the chromatin dynamics in CSCs and ESCs [96, 137]. Early and late cancer cells disseminate and reach secondary organs (bottom part of the panel), where they may acquire a distinct chromatin landscape that is likely to facilitate the maintenance of a dormancy program for an extended period of time. The transition of late cancer cells into dormancy involves the acquisition of repressive histone marks [6, 36–38] (H3K27/9me3 and H4K20; green hexagons) and upregulation of pluripotency genes (*SN*: *SOX2,* and *NANOG*) [6, 37]. Whether the same features happen in early DCCs remains to be seen. Therefore, late dormant DCCs upregulate pluripotency genes as seen in iPSCs/ESCs (**Panel A**), but differ from them because they present a global repressive chromatin. Investigations on the identification of exclusive open chromatin regions in DCCs and further comparison with open regions in iPSCs/ESCs sates may give insights on molecular mechanisms and how to apply this knowledge for therapy design. When specific signals trigger the exit of early and late DCCs from dormancy, they might adopt CSC-like properties, including an open chromatin, which favors metastatic outgrowth. Whether early or late dormant DCCs have the capacity to self-renew or whether they contain bivalent chromatin marks still needs to be investigated (*Note: this figure depicts active or repressive histone methylation modifications without details on DNA methylation*)

urokinase receptor and integrin beta 1, can regulate the dormant-to-proliferative switch by reducing p38 signaling [45] and activating myosin light chain kinase [46]. Moreover, the presence of DCCs may initiate a cross-talk with stromal cells that can eventually lead to a reciprocal reprogramming. In this regard, cancer associated fibroblast (CAF)-conditioned medium induced gene expression in tumor cells driven by the redistribution of the epigenetic reader methyl-CpG-binding domain protein 2 (MBD2) across the DNA methylated regions [47]. Importantly, CAF secreted factors can reprogram tumor cells and drive tumor progression [48]. Thus, programmed fibroblasts may induce the activation of DCCs by several mechanisms, including chromatin landscape modifications.

It has been proposed that DCCs displace hematopoietic stem cells (HSCs) and utilize the same specialized niche to survive at a low metabolic and quiescent state for longer periods of time [49]. Loss of autophagy is a common feature within an adverse aging microenvironment, which causes epigenetic rearrangements in HSCs leading to differentiation, exit from quiescence, and a reduced stemness capacity [50]. However, one-third of HSCs were still found to retain high levels of autophagy at these aging BM niches [50]. Interestingly, dormant tumor cells from a head and neck squamous cell carcinoma (HNSCC) model upregulated autophagy-related (ATG) genes [6, 51]. One can argue that DCCs might use these niches to maintain dormancy through the same mechanisms as those used by HSCs e.g., increased autophagy signaling pathway). However, within an aging microenvironment, the probabilities of reactivation increase, inducing changes in the epigenome of DCCs, and these changes might explain some cases of late metastatic relapses.

Bone morphogenetic proteins (BMPs) constitute the niche for adult stem cells and are known to induce quiescence in adult stem cells [52]. Similarly, BMPs can promote dormancy in cancer cells [53, 54]. BMP7, a transforming growth factor β (TGFβ) superfamily member, binds to BMPR2 receptor, activating p38 signaling and inducing dormancy of disseminated prostate tumor cells [54]. Interestingly, loss of BMPR2 correlates with bone metastases in prostate cancer patients [54]. Moreover, inhibition of BMP4 by the antagonist Coco (DAN5) was able to reactivate dormant 4T07 DCCs in lung tissues [55]. In a recent study, Adam et al. showed how external signals drive chromatin remodeling during hair follicle (HF) lineage determination. When blocking BMP signaling in mid-telogen (resting) phase, HFs displayed precocious proliferation, reduction in key bulge stem cell factors, and early signs of hair cycle activation in association with the acquisition of permissive chromatin patterns [56]. Thus, changes in the BMP activity may be able to rewire chromatin landscapes in DCCs to favor reawakening; however, the link with epigenetics needs to be further demonstrated.

1.2.2 Effect of Hypoxia on the Epigenome of DCCs

The modulation of DCCs and their epigenomes for the induction of dormancy could also be influenced and encoded by microenvironmental signals such as hypoxia before dissemination, in the primary sites [36]. Moreover, hypoxic conditions are known to control tumor gene expression by altering histone modifications [57]. Fluegen and co-authors [36] showed that hypoxic regions in the primary tumor sites induce histone 3 repressive marks and dormancy-related genes. This induces epigenetically predisposed tumor cells to remain in quiescence and survive treatments after dissemination to secondary sites. Consistently, the presence of dormancy signatures in primary breast cancer was an indicator of an extended overall patient survival [58]. Moreover, DCCs exposed to hypoxic environments at primary tumor sites were found to upregulate NR2F1 [36]. This orphan nuclear receptor and retinoic acid-responsive gene was shown to regulate the dormancy phase of residual post-surgery DCCs [6]. In addition, the identification of a global repressive chromatin characteristic in dormant cells was dependent on NR2F1 [6]. Whether the imposed repressive chromatin state in hypoxia is also NR2F1-dependent needs to be further determined. Similarly, another repressive mark on histone 4 lysine 20 (H4K20), induced by

Smyd5 methyltransferase, is known to regulate the dormancy phase of lung DCCs [38]. Further research is required to determine whether Smyd5 activity is induced by the hypoxic microenvironment inside primary tumors.

Metabolic stress due to hypoxia or fluctuations in the metabolite levels greatly influences the epigenetic control of gene expression in cancer. For instance, the metabolites S-adenosylmethionine (SAM) and α-ketoglutarate (αKG) affect histone methyltransferases and histone demethylases, respectively [57]. In glioblastomas, differentiated, weakly proliferative tumor regions showed elevated 4-hydroxybutyrate (GHB)/α-ketoglutarate ratios with scarce succinic semialdehyde dehydrogenase (SSADH) expression compared to regions characterized by non-differentiated, proliferative cells. GHB acts by inhibiting α-KG-dependent ten-eleven translocations (TET) enzyme activity, resulting in decreased levels of the 5-hydroxymethylcytosine epigenetic marks and inhibiting cell self-renewal, proliferation, and expression of stem markers [59]. Moreover, metabolic reprogramming has been shown to control transition from quiescence to proliferation in satellite cells (SC), which are skeletal muscle stem cells [60]. In growth-permissive conditions, SC experienced a metabolic shift from fatty acid and pyruvate oxidation to increased glycolysis and glutaminolysis. These metabolic modifications were associated with a decrease in the intracellular NAD^+/NADH ratio, which in turn reduced SIRT-1 mediated deacetylation of H4K16ac, activating the myogenic program [60]. Therefore, although not yet tested, epigenetic alterations controlled by metabolic reprogramming may induce dormancy, or exit from it, in DCCs.

1.2.3 Histone Variants and Long Non-coding RNAs Regulating DCC Quiescence

Histone variants and long non-coding RNAs may also play a role in regulating transitions between dormancy and reactivation. Reduced levels of macroH2A were found to be associated with melanoma aggressiveness [61]. Notably, macroH2A was found to be upregulated in dormant cells in an HNSCC model [51] and several studies have related it to both quiescence and senescence programs [62].

Studies from the Giancotti and Spector groups revealed that *MALAT1*, a long non-coding RNA, was required for the reactivation of dormant lung DCCs [38, 63] and favored tumor growth [64]. Non-coding RNAs, however, have emerged as important regulators of HSC self-renewal (such as HSC-enriched long non-coding RNAs named LncHSCs) and quiescence (such as H19) [65, 66] functions, as well as for the pluripotent state in ESCs [67]. Whether these long non-coding RNAs control quiescence and the self-renewal capacity of DCCs remains to be seen.

1.2.4 From Diapause to Cancer Dormancy

A developmental process that mimics cancer dormancy is called diapause [68]. Diapause is a regulated dormant state that allows animals and insects to survive adverse conditions and is terminated by endogenously controlled programs that are not well understood [68]. During the diapause phase of insects, changes in histone marks are known to affect transcriptional programs [69]. The use of azacytidine, a DNA methyltransferase inhibitor, or the knockdown of Dnmt1a (a DNA methyltransferase), reduced the ability of wasps (*Nasonia vitripennis*) to sense diapause-induced stimuli [70] suggesting that a controlled DNA methylation pattern is required to execute diapause-induced responses. Larvae of the sea snail (*Chymomyza costata*) that were reprogrammed to enter diapause showed a downregulation of Dpy-30 levels [71]. *Dpy-30* encodes a subunit of the Set1C-COM-PASS complex that methylates H3K4 [72]. Many of the genes regulated by DPY-30 are downregulated in diapause. Interestingly, hematopoietic stem cells and hematopoietic progenitor cells rely on the activity of DPY-30 to maintain proliferation and differentiation functions [73, 74]. Further research would be necessary to demonstrate whether DPY-30 plays a role in MRD, especially during the reactivation processes.

1.3 Epigenome Landscape in MRD After Therapeutic Treatments

1.3.1 Retinoic Acid and Inhibitors of Chromatin-Modifying Enzymes Favor Tumor Latency

Current therapeutic drugs might modulate the epigenome of residual cancer cells and consequently affect their cell fate. For instance, breast cancer cells treated *ex vivo* with transient nanomolar doses of decitabine, a DNA demethylating agent, were found to undergo extended latency when injected into mice [75]. This response was associated with a decrease in $CD44^{+}/ALDH^{+}$ levels, downregulation of DNA (cytosine-5)-methyltransferase 1 (DNMT1), and forkhead box protein M1 (FOXM1), and upregulation of retinoic acid receptor beta (RARβ) and cyclin-dependent kinase inhibitor 1A (CDKN1A) [75]. Similar changes in the regulation of these genes were identified in a dormancy model of HNSCC [6, 51]. Furthermore, a combination treatment with azacytidine and all-trans retinoic acid (atRA) induced HNSCC tumor cells with genetic aberrations to enter a long-lasting dormancy *in vivo* [6]. These reprogrammed tumor cells showed an upregulation of the NR2F1-dependent dormancy program, which included the RARβ and cell cycle inhibitors [6]. NR2F1 is expressed during human embryonic stem cell differentiation where it was found to regulate cell lineage commitment genes (such as SRY [sex determining region Y]- box 9 (*SOX9*)) by binding to enhancers [76]. NR2F1 has been shown to be required not only to maintain the dormancy program in HNSCC, but also to control key genes responsive to the retinoic acid signaling, including *SOX9* and *RARβ* [6]. In addition, the conversion of tumor cells into dormant cells by azacytidine and atRA was accompanied by a global gain in repressive histone marks ($H3K9me3^{high}/H3K27me3^{high}/H3K4me3^{low}$), which were partially dependent on NR2F1 [6]. These results indicated that NR2F1 might regulate histone modifications, but further analysis is required. Moreover, azacytidine treatment has been found to cause changes in histone marks [77]. atRA also modulated DNA methylation [78] and histone marks [79]. Overall, these results suggest that the modulation of DNA methylation and histone marks can induce gene expression associated with dormancy.

Treatment of uveal melanoma cells with histone deacetylase (HDAC) inhibitors caused cell growth arrest by epigenetic deregulation and signs of differentiation [80]. A recent study revealed that a novel HDAC inhibitor (JSL-1) induced apoptosis in uveal melanoma cells and impaired the self-renewal capacity of stem cell-like cells [81].

Resistance to drugs also involves chromatin and histone activity alterations. The H3K4 demethylase lysine-specific demethylase 5A (KDM5A) was shown to be required for the survival of quiescent drug-tolerant subpopulations or "persisters" upon treatment with epithelial growth factor receptor (EGFR) tyrosine kinase inhibitors in several types of cancer cell lines [82]. Moreover, Trichostatin A, an HDAC inhibitor, killed the quiescent persisters [82]. Based on these studies, it remains to be shown in animal models whether modulations of chromatin-remodeling enzymes might affect the survival of residual DCCs.

1.3.2 Epigenetic Effects of Tamoxifen in MRD

Tamoxifen has been the standard treatment for women with ER-positive cancers [83]. Tamoxifen binds to estrogen receptor (ER) and inhibits ER-regulated gene transcription in ER-positive breast cancer cells. One mechanism of action for tamoxifen is the recruitment of histone deacetylases (HDACs) that induce H3 and H4 deacetylation, for example, in pS2 and c-myc gene promoters [84]. In resistant breast cancer cells, tamoxifen also promotes DNA methylation in the downstream targets of estrogen, pS2, the progesterone receptor (PGR), and growth differentiation factor 15 (GDf15). Treatment with azacytidine and estradiol (E2) could induce an anti-proliferative effect in tamoxifen-resistant cells [85]. This research suggests that tamoxifen can induce silencing of ER-regulated gene promoters and that reactivation of specific genes by using epigenetic drugs may be beneficial in tamoxifen-resistant patients. Notably, early-stage ER-positive breast cancer patients that respond to tamoxifen may undergo remission for extended periods of time, sometimes decades, before they relapse. In these cases, elucidating whether tamoxifen modifies chromatin in residual cancer cells in a way that keeps them dormant may create new therapeutic interventions to extend asymptomatic MRD in other types of cancers.

The resistance of breast cancer cells to tamoxifen relies on epigenetic changes induced by their microenvironment. For instance, the inflammatory cytokine IL-1β, secreted mainly by macrophages localized in the tumor microenvironment, induces methylation of the ER gene through the activation of specific effectors, including the IL-1β/IL-1RI/β-catenin signaling pathway and Twist-related protein 1 (TWIST1) [86]. In one study, dormant early DCCs in the lung derived from early lesions in the mammary epithelium showed TWIST upregulation and this was reverted in later stages of DCCs [30]. Interestingly, early precursors of breast DCCs interacted with macrophages at mammary glands, increasing the β-catenin signal and favoring early dissemination [30]. Based on this evidence, one may hypothesize that early DCCs resist tamoxifen adjuvant treatments and become the main sources for later metastatic outgrowth. However, further investigations demonstrating the levels of ER and its activity in early DCCs need to be done.

1.3.3 BRAF and Chromatin Remodeler Inhibitors During MRD in Melanoma

In melanoma, B-Raf proto-oncogene (BRAF) V600E mutations have been found to epigenetically silence specific genes that influence disease progression [87]. BRAF V600E increases expression levels of MAF BZIP transcription factor G (MAFG), a transcriptional repressor resulting in hypermethylation and epigenetic silencing [87]. Another mechanism of BRAF V600E-driven DNA hypermethylation is via the upregulation of DNMT1 [88]. Recently, a phase-I clinical study tested the action of vemurafenib (a BRAF inhibitor) in combination with a low dose of decitabine, and showed overall clinical benefits in 79% of the patients [89]. Interestingly, 48 days of the exposure of melanoma cells *in vitro* to vemurafenib in combination with decitabine, maintained DNMT1 depletion whereas emergent resistant clones to vemurafenib alone regained expression of DNMT1. Although no significant short-term apoptosis was observed *in vitro*, further studies are needed to fully elucidate the effects of this drug combination in cancer patients [89]. In contrast to the effects of decitabine, the Bromodomain and Extra-Terminal domain (BET) inhibitor JQ1 interacted synergistically with vemurafenib *in vitro*, inducing significantly more apoptosis in melanoma cells [90]. In a xenograft mouse model of BRAF mutant melanoma, this combination therapy impaired tumor growth and extended survival in mice compared to the single agent groups [90]. These findings imply that targeting epigenetic modulators with a combination of BRAF inhibitors is a promising therapeutic approach to extend MRD in melanoma. However, whether this combination of drugs controls the quiescence phase or affects the survival of residual tumor cells has yet to be determined.

1.3.4 Androgen-Deprivation Therapy May Induce a Dormancy Program in Prostate Cancer

In prostate cancer, androgen-induced androgen receptor (AR) binding and HDAC-mediated deacetylation at the secreted protein acidic and rich in cysteine (SPARC)-like 1 (SPARCL1) locus suppresses its expression and facilitates disease progression [91]. Moreover, this response could be attenuated using AR antagonists or HDAC inhibitors [91]. Interestingly, SPARCL1 is induced by a p38-dependent dormancy signaling in HNSCC cells, suggesting a potential role during the quiescence phase of prostate cancer [51]. SPARC, a paralog of SPARCL1 that shares many structural similarities with SPARCL1, was recently shown to induce the dormancy in prostate cancer cells in bones [53]. Similar to *SPARCL1*, *SPARC* was epigenetically silenced in aggressive prostate cancer cells but was re-expressed in indolent cancer cells in the bone and was responsible for BMP7 induction [53]. Androgen depletion therapy (ADT) induces remission in approximately 90% of prostate cancer patients. Analysis of residual tumor cells upon ADT demonstrated an upregulation of *NR2F1* [92]. Computational analyses of the dormancy signature discovered in HNSCC showed that NR2F1 positively regulated SPARC [51], although the mechanisms of regulation are not yet known. Importantly, NR2F1 was found to be responsible for the repressive chromatin state acquired by dormant cancer cells [6]. However, 2–3 years after ADT, tumor regrowth [93] was observed, probably accompanied by the downregulation of these dormancy genes.

Thus, blocking androgen functions might induce a dormancy-like response, which could be further sustained beyond the 2–3 year mark by a dormancy-inducing treatment that preserves the upregulation of *NR2F1*, *SPARC*, and *BMP7* together with a dormancy-like chromatin landscape.

In summary, these findings suggest that current therapies could induce epigenetic modifications that prevent pro-growth genetic changes in early and late DCCs and regulate their cell fate. Epigenetic therapies could therefore help prevent DCCs from becoming deadly metastases. Moreover, the identification of epigenetic markers in DCCs could be useful in monitoring responses to epigenetic treatment in cancer patients.

1.4 Pluripotency Genes: Similarities Between DCCs and Stem Cells

ESCs have to fulfill two characteristics: self-renewal and the ability to differentiate into different cell types. These cells rely on epigenetic landscapes in order to maintain chromatin in such a way that they can self-renew or differentiate [94] (Fig. 1.1). Epigenome flexibility is also illustrated by the reprogramming of somatic cells into induced pluripotent stem cells (iPSCs), generated by the expression of four defined transcription factors (TFs), OCT3/4, SOX2, KLF4, and MYC (OSKM) [95]. iPSCs, like ESCs, are capable of robust proliferation and differentiation into many different cell types. During the reprogramming of somatic cells to iPSCs, MYC was found associated to the promoters of its targets bearing active histone marks and it interacted with histone acetyltransferase complexes [96, 97]. In the dormancy model of HNSCCs, the *MYC* node is deactivated when cells become dormant and acquire repressive chromatin [51]. This is in agreement with the study by Scognamiglio et al. that showed that deactivation of *MYC* induced reversible dormancy in blastocysts without affecting the pluripotency [98]. Moreover, dormant cancer cells upregulated the levels of pluripotency genes *SOX2* and *NANOG* [6, 37] (Fig. 1.1), which are also activated in iPSCs [97]. Dormancy induced by the downregulation of *MERTK* in prostate cancer also induced increased expression from *SOX2* and *NANOG*, among other dormancy genes [37]. SOX2 has been shown to maintain the quiescence of inner pillar cells via p27 [99] as well as in retinal Muller glia [100]. The function of these pluripotency genes in dormant DCCs, whether it is maintenance of quiescence, survival signals, and/or latent tumor-initiating capacity, remains to be elucidated. Interestingly, it has recently been reported that patients with SOX2-negative hypopharyngeal squamous cell carcinoma (HPSCC) had significantly worse overall and recurrence-free survival rates than those with SOX2-positive cancer. In this study, *SOX2* was epigenetically silenced by promoter hypermethylation [101].

Notably, several characteristics are shared between adult stem cells and DCCs. Approximately 60% of genes upregulated in adult stem cells were also induced in dormant tumor cells [102]. Methylation status regulates quiescence of adult stem cells [103]. For instance, the H4K20 dimethyltransferase SUV420H1 was responsible for skeletal muscle stem cell quiescence by blocking the expression of the specific regulator MYOD, which marks myogenic commitment [104]. As mentioned before in Section 1, another H4K20 methyltransferase, SMYD5, was found to be required for the dormancy phase of DCCs [38]. Dormant HSCs maintain a high expression of a retinoic acid program [105] that has been previously found in dormant DCCs [6]. Moreover, the perivascular niche is responsible for DCC dormancy, like in stem cell biology [106]. All these data point to similarities between dormant DCCs and some features of embryonic and adult stem cells. However, several questions remain: Are dormant DCCs able to self-renew, and are they able to undergo terminal differentiation? The latter is an interesting idea. In an attempt to supersede the genetic and epigenetic alterations that blocked differentiation and promoted tumorigenesis in sarcomas, Zhang et al. published a pluripotency-based reprogramming strategy [107]. In their study, they showed that directly reprogrammed sarcomas (but not parental cells) had a broad differentiation potential and could terminally execute normal cellular differentiation pathways with the consequent loss of tumorigenicity. Along these lines, it would be interesting to test if the tumor cells reprogrammed into dormancy with low doses of azacytidine and retinoic acid (mentioned in 2.1. and [6]), which upregulated *SOX2* and *NANOG*, are more susceptible to undergo terminal differentiation in the presence of certain stimuli. Further investigation would be needed to address these questions and to reveal pathways to intervene and extend MRD by targeting DCCs.

1.5 DCCs, CCCs, and Cell-Free DNA as Surrogates for MRD Monitoring

Numerous clinical studies have analyzed the presence of DCCs, CCCs, and more recently, cell-free DNA as measurable biomarkers of MRD to determine disease progression and response to therapies. In 1998, a clinical trial that included neoadjuvant ADT was carried out in 101 patients with clinically localized prostate cancer [108]. After 6 months of treatment, the number of BM DCCs decreased by 86% in patients with an initial detection of DCCs. Similarly, adjuvant and neoadjuvant therapies in early breast cancer patients were found to reduce the number of BM DCCs, but not completely eliminate them [109, 110]. Importantly, in most of these studies, the initial counting of BM DCCs was carried out in the presence of a primary tumor. Thus, the regression of a primary tumor itself could account for the decrease observed in the number of DCCs at the end of treatment and may not reflect the actual elimination of DCCs. Therefore, one way to study DCCs as tools to monitor MRD is by using sequential bone marrow aspirates after primary tumor removal. For instance, Naume and co-authors reported on secondary treatments in early breast cancer patients (patients with no metastasis) who were positive for BM DCCs after receiving anthracycline-containing treatments [111]. The results showed that the absence of DCCs after secondary treatment with docetaxel correlated with survival in a follow-up study of 6 years. However, 8.8% of the patients negative for DCCs after docetaxel still relapsed. Moreover, the presence of DCCs is not always followed by metastasis [6, 10]. As such, the molecular analysis of DCC populations is of clinical relevance in order to avoid unnecessary treatments and to be able to distinguish patients who are at risk of imminent relapses from those who have stable disease.

Another promising therapy targeting DCCs is the use of bisphosphonate (BP). There are various preclinical studies showing the effect of using bisphosphonate as an adjuvant during therapy on improving disease-free survival and overall survival in early breast cancer patients (with no metastases) who were initially positive for DCCs, and on the elimination of BM DCCs [23, 112, 113]. Animal model-based studies have shown that BP inhibits invasion, proliferation, angiogenesis, and osteoclast-mediated bone reabsorption, and it also induces apoptosis of cancer cells [113]. These results argue that the use of BP may prevent metastatic relapses. It remains to be seen which population of DCCs BP targets and the mechanisms of action involved, whether it has a direct effect on DCCs, or on the modifications of the niche, or both.

Over the last 50 years of cancer research, we have learned that in order to reduce cancer mortality rates we need to focus on treating metastasis. Owing to the fact that obtaining repeated samples from metastatic sites is not possible, the study of CCCs and circulating DNA during MRD has become of extreme importance. In 1869, Ashworth discovered the first CCCs [114] in blood samples of cancer patients. Since then, several studies have used peripheral blood to detect CCCs in non-metastatic cancer patients and in advanced-stage cancer patients, providing ways to monitor disease progression and to predict the patient's response to therapy [115–117]. The latest advances in the field of liquid biopsy have introduced the leukapheresis (LA) method to isolate higher volumes of blood allowing for the detection of CCCs even in M0-stage patients [118]. Moreover, CCCs have been found and analyzed in breast cancer patients in clinical dormancy without any evidence of disease recurrence [17, 18, 118]. Patients in clinical dormancy showed a higher percentage of CCCs negative for the proliferation marker Ki67 and positive for the apoptotic marker M30 compared to patients with relapses [18]. The explanation for this observation needs further investigation. Although enumeration of CCCs before and after therapy has been used as a prognosis marker [115–117], only a few studies have given a comprehensive molecular analysis of CCCs, which currently limits their use as surrogates for MRD monitoring and selection of targeted drugs. Several studies support the idea that the correlation between DCCs and CCCs is low, arguing that there is a different clinical relevance in their shared characteristics [119]. Molloy and co-authors showed that CCC detection in early breast cancer was a better prognostic marker for an ER-negative subset than the presence of DCCs [120]. On the other hand, the detection of BM DCCs resulted in a superior prognostic value than did CCCs in other breast cancer studies [121, 122]. All these findings indicate that CCCs and DCCs reflect different aspects of the disease, including the tumor subtypes and time for metastasis relapses. One important point worth noting is the utilization of different methodologies to detect CCCs and DCCs, which could affect the interpretation of the results and contribute to conflicting results.

Lastly, it is known that circulating cancer cell-free DNA can be detected during MRD [123]. This fragmented tumor DNA most likely originates from cancer cells that underwent necrosis or apoptosis. In follow-up studies, up to 12 years later, patients' cell-free DNA was found to have copy number variations (CNV) and loss of heterozygosity (LOH), suggesting that a reservoir of tumorigenic cells might enter the circulation during MRD [123]. The question that needs to be addressed now is whether the genomic and/or epigenetic profile of cell-free DNA could help to predict if a patient will relapse and if so, when. In December 2017, Widschwendter et al. published a study showing the predictive value of DNA methylation patterns in serum DNA [124]. They were able to clinically validate a specific methylated region located within the *GP5* gene that predicted fatal breast cancer up to one year in advance.

1.6 Concluding Remarks

In the last 20 years, significant progress has been made in the field of cancer epigenetics. Several studies have defined epigenetic profiles in primary tumors in order to predict patient outcomes. Cutaneous melanoma and nasopharyngeal carcinoma patients with a high DNA methylation profile have been found to have the worst prognoses [125, 126]. In addition, epigenetic alterations at gene enhancer elements have been found to drive a unique signature that promotes colon carcinogenesis [127]. Here, the authors proposed that the epigenetic signature in enhancers is less heterogeneous between different colorectal cancer samples than mutations in driver genes. Reduction in the levels of 5hmC methylation in melanoma [128] and glioblastomas was also associated with a poor outcome [129]. Studies have revealed a correlation between 5hmC patterns and genome integrity [130]. Finally, DNA methylation patterns have been found to help in identifying the tumor type origin of CUP [131].

We have learned that the prevention of the formation of metastasis is a more effective way to extend patient survival than is the treatment of detectable metastases. Our understanding of the biology of MRD may be improved by studying DCCs, CCCs, and cell-free DNA in greater depth. The identification of DNA methylation and histone post-translational modifications in DCCs (early and late), CCCs, and cell-free DNA might provide information regarding disease progression and may allow us to monitor patient responses to treatment.

As our understanding of DCC and CCC biology continues to grow (Table 1.1), the next step is to find a way to turn these findings into new prognostic and therapeutic approaches for MRD. The evidence of early cancer cell dissemination as another source for dormant DCCs emphasizes the heterogeneity of MRD. A deeper understanding of the crosstalk between niche signals and the DCC dormancy program is paramount for the development of potential new treatment strategies. Moreover, DCCs could be epigenetically targeted to induce or maintain a dormancy program, or to be eliminated altogether. We also need to identify therapeutic markers during MRD (Table 1.1) (i.e., DCC biomarkers that can predict early vs. late relapses) that physicians can use to select the right treatment and to monitor drug responses in patients.

We still have a lot to learn in terms of how to design new therapies and when to combine treatments that eliminate primary tumor masses with those that target MRD. The ongoing investigations in the field of dormancy, the clinical therapeutic evidence during MRD, and the latest advances in genetic and epigenetic technologies will certainly help researchers achieve some of these goals.

Acknowledgments M.S.S. was supported by the Schneider-Lesser Foundation Fellow Award, the Melanoma Research Alliance (MRA), the NCI Transition Career Development Award (K22) (22CA 201054), and the Susan G. Komen Career Catalyst Research Grants (basic/translational and clinical research) CCR17483357.

Disclosure Statement The authors are not aware of any affiliations, memberships, funding, or financial holdings that might be perceived as affecting the objectivity of this review.

Table 1.1 Comparison of previously described [4] and recently identified novel DCC dormancy markers

Type of cancer	DCCs and niche markers (2014) [4]	DCCs and niche markers (present)
Prostate, breast, and HNSCC	DCC: TGFβR3 ↑, BMPR2 ↑, P-ERK ↓, P-p38 ↑, GRP78 ↑	DCC: Same as in 2014 plus: NR2F1 [6, 36, 37]↑, SOX2 [6, 37] ↑, NANOG [6, 37] ↑, SPARC [53]↑, H3K9me3 [6, 37]↑, H3K27me3 [6, 36, 37] ↑, Smyd5 [38] ↑, DEC2 [36]↑, MERTK [37]↓, MALAT↓ [38]
	Niche: TGFβ2 ↑, BMP4 and BMP7 ↑	Niche: SDF-1 [138]↑

These biomarkers could be used during MRD to determine the status of DCCs after neo-adjuvant and adjuvant treatments, to select suitable dormancy-inducing drugs, and to determine long-lasting responses to treatment over the years

References

1. Howlader N, Noone AM, Krapcho M, Miller D, Bishop K, Kosary CL, Yu M, Ruhl J, Tatalovich Z, Mariotto A, Lewis DR, Chen HS, Feuer EJ, Cronin KA (2017) SEER Cancer Statistics Review (CSR) 1975-2014. National Cancer Institute, Bethesda, MD
2. Aguirre-Ghiso JA (2007) Models, mechanisms and clinical evidence for cancer dormancy. Nat Rev Cancer 7:834–846
3. Dasgupta A, Lim AR, Ghajar CM (2017) Circulating and disseminated tumor cells: harbingers or initiators of metastasis? Mol Oncol 11:40–61. https://doi.org/10.1002/1878-0261.12022
4. Sosa MS, Bragado P, Aguirre-Ghiso JA (2014) Mechanisms of disseminated cancer cell dormancy: an awakening field. Nat Rev Cancer 14:611–622. https://doi.org/10.1038/nrc3793
5. Giancotti FG (2013) Mechanisms governing metastatic dormancy and reactivation. Cell 155:750–764. https://doi.org/10.1016/j.cell.2013.10.029

6. Sosa MS et al (2015) NR2F1 controls tumour cell dormancy via SOX9- and RARbeta-driven quiescence programmes. Nat Commun 6:6170. https://doi.org/10.1038/ncomms7170
7. Bragado P et al (2013) TGF-beta2 dictates disseminated tumour cell fate in target organs through TGF-beta-RIII and p38alpha/beta signalling. Nat Cell Biol 15:1351–1361. https://doi.org/10.1038/ncb2861
8. Noltenius C, Noltenius H (1985) Dormant tumor cells in liver and brain. An autopsy study on metastasizing tumors. Pathol Res Pract 179:504–511. https://doi.org/10.1016/S0344-0338(85)80191-6
9. Braun S et al (2000) Cytokeratin-positive cells in the bone marrow and survival of patients with stage I, II, or III breast cancer. N Engl J Med 342:525–533
10. Chery L et al (2014) Characterization of single disseminated prostate cancer cells reveals tumor cell heterogeneity and identifies dormancy associated pathways. Oncotarget 5:9939–9951
11. Schardt JA et al (2005) Genomic analysis of single cytokeratin-positive cells from bone marrow reveals early mutational events in breast cancer. Cancer Cell 8:227–239
12. Rud AK et al (2016) Detection of disseminated tumor cells in lymph nodes from patients with early stage non-small cell lung cancer. Diagn Pathol 11:50. https://doi.org/10.1186/s13000-016-0504-4
13. Dhayat S et al (2012) Prognostic significance of EpCAM-positive disseminated tumor cells in rectal cancer patients with stage I disease. Am J Surg Pathol 36:1809–1816. https://doi.org/10.1097/PAS.0b013e318265288c
14. Sproll C et al (2017) Immunohistochemical detection of lymph node-DTCs in patients with node-negative HNSCC. Int J Cancer 140:2112–2124. https://doi.org/10.1002/ijc.30617
15. Linde N, Fluegen G, Aguirre-Ghiso JA (2016) The relationship between dormant cancer cells and their microenvironment. Adv Cancer Res 132:45–71. https://doi.org/10.1016/bs.acr.2016.07.002
16. Ptak GE et al (2012) Embryonic diapause is conserved across mammals. PLoS ONE 7:e33027. https://doi.org/10.1371/journal.pone.0033027
17. Meng S et al (2004) Circulating tumor cells in patients with breast cancer dormancy. Clin Cancer Res 10:8152–8162
18. Spiliotaki M et al (2014) Evaluation of proliferation and apoptosis markers in circulating tumor cells of women with early breast cancer who are candidates for tumor dormancy. Breast Cancer Res 16:485. https://doi.org/10.1186/s13058-014-0485-8
19. Strauss DC, Thomas JM (2010) Transmission of donor melanoma by organ transplantation. Lancet Oncol 11:790–796. https://doi.org/10.1016/S1470-2045(10)70024-3
20. Friberg S, Nystrom A (2015) Cancer metastases: early dissemination and late recurrences. Cancer Growth Meta 8:43–49. https://doi.org/10.4137/CGM.S31244
21. Braun S et al (2005) A pooled analysis of bone marrow micrometastasis in breast cancer. N Engl J Med 353:793–802
22. Janni W et al (2011) Persistence of disseminated tumor cells in the bone marrow of breast cancer patients predicts increased risk for relapse--a European pooled analysis. Clin Cancer Res Off J Am Assoc Cancer Res 17:2967–2976. https://doi.org/10.1158/1078-0432.CCR-10-2515
23. Hartkopf AD et al (2014) Prognostic relevance of disseminated tumour cells from the bone marrow of early stage breast cancer patients - results from a large single-centre analysis. Eur J Cancer 50:2550–2559. https://doi.org/10.1016/j.ejca.2014.06.025
24. Sosa MS (2016) Dormancy programs as emerging antimetastasis therapeutic alternatives. Mol Cell Oncol 3:e1029062. https://doi.org/10.1080/23723556.2015.1029062
25. Sanger N et al (2011) Disseminated tumor cells in the bone marrow of patients with ductal carcinoma in situ. Int J Cancer 129:2522–2526. https://doi.org/10.1002/ijc.25895
26. Deryugina EI, Kiosses WB (2017) Intratumoral Cancer Cell Intravasation Can Occur Independent of Invasion into the Adjacent Stroma. Cell Rep 19:601–616. https://doi.org/10.1016/j.celrep.2017.03.064
27. Klein CA (2009) Parallel progression of primary tumours and metastases. Nat Rev Cancer 9:302–312
28. Schmidt-Kittler O et al (2003) From latent disseminated cells to overt metastasis: genetic analysis of systemic breast cancer progression. Proc Natl Acad Sci U S A 100:7737–7742
29. Werner-Klein M et al (2018) Genetic alterations driving metastatic colony formation are acquired outside of the primary tumour in melanoma. Nat Commun 9:595. https://doi.org/10.1038/s41467-017-02674-y
30. Linde N et al (2018) Macrophages orchestrate breast cancer early dissemination and metastasis. Nat Commun 9:21. https://doi.org/10.1038/s41467-017-02481-5
31. Harper K et al (2016) Mechanism of early dissemination and metastasis in Her2(+) mammary cancer. Nature. https://doi.org/10.1038/nature20609
32. Hosseini H et al (2016) Early dissemination seeds metastasis in breast cancer. Nature. https://doi.org/10.1038/nature20785
33. Husemann Y et al (2008) Systemic spread is an early step in breast cancer. Cancer Cell 13:58–68
34. Rhim AD et al (2012) EMT and dissemination precede pancreatic tumor formation. Cell 148:349–361. https://doi.org/10.1016/j.cell.2011.11.025
35. Narod SA, Iqbal J, Giannakeas V, Sopik V, Sun P (2015) Breast cancer mortality after a diagnosis of ductal carcinoma *in situ*. JAMA Oncol 1:888–896. https://doi.org/10.1001/jamaoncol.2015.2510
36. Fluegen G et al (2017) Phenotypic heterogeneity of disseminated tumour cells is preset by primary tumour hypoxic microenvironments. Nat Cell Biol 19:120–132. https://doi.org/10.1038/ncb3465

37. Cackowski FC et al (2017) Mer tyrosine kinase regulates disseminated prostate cancer cellular dormancy. J Cell Biochem 118:891–902. https://doi.org/10.1002/jcb.25768
38. Gao H et al (2014) Forward genetic screens in mice uncover mediators and suppressors of metastatic reactivation. Proc Natl Acad Sci U S A 111:16532–16537. https://doi.org/10.1073/pnas.1403234111
39. Lu P, Weaver VM, Werb Z (2012) The extracellular matrix: a dynamic niche in cancer progression. J Cell Biol 196:395–406. https://doi.org/10.1083/jcb.201102147
40. Le Beyec J et al (2007) Cell shape regulates global histone acetylation in human mammary epithelial cells. Exp Cell Res 313:3066–3075. https://doi.org/10.1016/j.yexcr.2007.04.022
41. Spencer VA, Xu R, Bissell MJ (2007) Extracellular matrix, nuclear and chromatin structure, and gene expression in normal tissues and malignant tumors: a work in progress. Adv Cancer Res 97:275–294. https://doi.org/10.1016/S0065-230X(06)97012-2
42. Xu R, Spencer VA, Bissell MJ (2007) Extracellular matrix-regulated gene expression requires cooperation of SWI/SNF and transcription factors. J Biol Chem 282:14992–14999. https://doi.org/10.1074/jbc.M610316200
43. Kaukonen R et al (2016) Normal stroma suppresses cancer cell proliferation via mechanosensitive regulation of JMJD1a-mediated transcription. Nat Commun 7:12237. https://doi.org/10.1038/ncomms12237
44. Kaplan RN et al (2005) VEGFR1-positive haematopoietic bone marrow progenitors initiate the pre-metastatic niche. Nature 438:820–827. https://doi.org/10.1038/nature04186
45. Aguirre-Ghiso JA, Liu D, Mignatti A, Kovalski K, Ossowski L (2001) Urokinase receptor and fibronectin regulate the ERK(MAPK) to p38(MAPK) activity ratios that determine carcinoma cell proliferation or dormancy in vivo. Mol Biol Cell 12:863–879
46. Barkan D et al (2008) Inhibition of metastatic outgrowth from single dormant tumor cells by targeting the cytoskeleton. Cancer Res 68:6241–6250. https://doi.org/10.1158/0008-5472.CAN-07-6849
47. Mathot P et al (2017) DNA methylation signal has a major role in the response of human breast cancer cells to the microenvironment. Oncogene 6:e390. https://doi.org/10.1038/oncsis.2017.88
48. Luo H, Tu G, Liu Z, Liu M (2015) Cancer-associated fibroblasts: a multifaceted driver of breast cancer progression. Cancer Lett 361:155–163. https://doi.org/10.1016/j.canlet.2015.02.018
49. Shiozawa Y et al (2011) Human prostate cancer metastases target the hematopoietic stem cell niche to establish footholds in mouse bone marrow. J Clin Invest 121:1298–1312. https://doi.org/10.1172/JCI43414
50. Ho TT et al (2017) Autophagy maintains the metabolism and function of young and old stem cells. Nature 543:205–210. https://doi.org/10.1038/nature21388
51. Adam AP et al (2009) Computational identification of a p38SAPK-regulated transcription factor network required for tumor cell quiescence. Cancer Res 69:5664–5672
52. Hsu YC, Pasolli HA, Fuchs E (2011) Dynamics between stem cells, niche, and progeny in the hair follicle. Cell 144:92–105. https://doi.org/10.1016/j.cell.2010.11.049
53. Sharma S et al (2016) Secreted Protein Acidic and Rich in Cysteine (SPARC) mediates metastatic dormancy of prostate cancer in bone. J Biol Chem 291:19351–19363. https://doi.org/10.1074/jbc.M116.737379
54. Kobayashi A et al (2011) Bone morphogenetic protein 7 in dormancy and metastasis of prostate cancer stem-like cells in bone. J Exp Med 208:2641–2655. https://doi.org/10.1084/jem.20110840
55. Gao H et al (2012) The BMP inhibitor coco reactivates breast cancer cells at lung metastatic sites. Cell 150:764–779. https://doi.org/10.1016/j.cell.2012.06.035
56. Adam RC et al (2018) Temporal layering of signaling effectors drives chromatin remodeling during hair follicle stem cell lineage progression. Cell Stem Cell. https://doi.org/10.1016/j.stem.2017.12.004
57. Tran TQ, Lowman XH, Kong M (2017) Molecular pathways: metabolic control of histone methylation and gene expression in cancer. Clin Cancer Res 23:4004–4009. https://doi.org/10.1158/1078-0432.CCR-16-2506
58. Kim RS et al (2012) Dormancy signatures and metastasis in estrogen receptor positive and negative breast cancer. PLoS ONE 7:e35569. https://doi.org/10.1371/journal.pone.0035569
59. El-Habr EA et al (2017) A driver role for GABA metabolism in controlling stem and proliferative cell state through GHB production in glioma. Acta Neuropathol 133:645–660. https://doi.org/10.1007/s00401-016-1659-5
60. Ryall JG et al (2015) The NAD(+)-dependent SIRT1 deacetylase translates a metabolic switch into regulatory epigenetics in skeletal muscle stem cells. Cell Stem Cell 16:171–183. https://doi.org/10.1016/j.stem.2014.12.004
61. Kapoor A et al (2010) The histone variant macroH2A suppresses melanoma progression through regulation of CDK8. Nature 468:1105–1109. https://doi.org/10.1038/nature09590
62. Chen H et al (2015) MacroH2A1 and ATM play opposing roles in paracrine senescence and the senescence-associated secretory phenotype. Mol Cell 59:719–731. https://doi.org/10.1016/j.molcel.2015.07.011
63. Arun G et al (2016) Differentiation of mammary tumors and reduction in metastasis upon Malat1 lncRNA loss. Genes Dev 30:34–51. https://doi.org/10.1101/gad.270959.115

64. Gutschner T et al (2013) The noncoding RNA MALAT1 is a critical regulator of the metastasis phenotype of lung cancer cells. Cancer Res 73:1180–1189. https://doi.org/10.1158/0008-5472.CAN-12-2850
65. Luo M et al (2015) Long non-coding RNAs control hematopoietic stem cell function. Cell Stem Cell 16:426–438. https://doi.org/10.1016/j.stem.2015.02.002
66. Venkatraman A et al (2013) Maternal imprinting at the H19-Igf2 locus maintains adult haematopoietic stem cell quiescence. Nature 500:345–349. https://doi.org/10.1038/nature12303
67. Guttman M, Rinn JL (2012) Modular regulatory principles of large non-coding RNAs. Nature 482:339–346. https://doi.org/10.1038/nature10887
68. Kostal V (2006) Eco-physiological phases of insect diapause. J Insect Physiol 52:113–127. https://doi.org/10.1016/j.jinsphys.2005.09.008
69. Reynolds JA, Bautista-Jimenez R, Denlinger DL (2016) Changes in histone acetylation as potential mediators of pupal diapause in the flesh fly, Sarcophaga bullata. Insect Biochem Mol Biol 76:29–37. https://doi.org/10.1016/j.ibmb.2016.06.012
70. Pegoraro M, Bafna A, Davies NJ, Shuker DM, Tauber E (2016) DNA methylation changes induced by long and short photoperiods in Nasonia. Genome Res 26:203–210. https://doi.org/10.1101/gr.196204.115
71. Poupardin R et al (2015) Early transcriptional events linked to induction of diapause revealed by RNAseq in larvae of drosophilid fly, Chymomyza costata. BMC Genomics 16:720. https://doi.org/10.1186/s12864-015-1907-4
72. Ardehali MB et al (2011) Drosophila Set1 is the major histone H3 lysine 4 trimethyltransferase with role in transcription. EMBO J 30:2817–2828. https://doi.org/10.1038/emboj.2011.194
73. Yang Z et al (2014) The DPY30 subunit in SET1/MLL complexes regulates the proliferation and differentiation of hematopoietic progenitor cells. Blood 124:2025–2033. https://doi.org/10.1182/blood-2014-01-549220
74. Yang Z, Shah K, Khodadadi-Jamayran A, Jiang H (2016) Dpy30 is critical for maintaining the identity and function of adult hematopoietic stem cells. J Exp Med 213:2349–2364. https://doi.org/10.1084/jem.20160185
75. Tsai HC et al (2012) Transient low doses of DNA-demethylating agents exert durable antitumor effects on hematological and epithelial tumor cells. Cancer Cell 21:430–446. https://doi.org/10.1016/j.ccr.2011.12.029
76. Rada-Iglesias A et al (2012) Epigenomic annotation of enhancers predicts transcriptional regulators of human neural crest. Cell Stem Cell 11:633–648. https://doi.org/10.1016/j.stem.2012.07.006
77. Komashko VM, Farnham PJ (2010) 5-azacytidine treatment reorganizes genomic histone modification patterns. Epigenetics 5:229–240
78. Miftakhova R et al (2012) DNA methylation in ATRA-treated leukemia cell lines lacking a PML-RAR chromosome translocation. Anticancer Res 32:4715–4722
79. Phipps SM, Love WK, White T, Andrews LG, Tollefsbol TO (2009) Retinoid-induced histone deacetylation inhibits telomerase activity in estrogen receptor-negative breast cancer cells. Anticancer Res 29:4959–4964
80. Landreville S et al (2012) Histone deacetylase inhibitors induce growth arrest and differentiation in uveal melanoma. Clin Cancer Res 18:408–416. https://doi.org/10.1158/1078-0432.CCR-11-0946
81. Wang Y, Liu M, Jin Y, Jiang S, Pan J (2017) In vitro and in vivo anti-uveal melanoma activity of JSL-1, a novel HDAC inhibitor. Cancer Lett 400:47–60. https://doi.org/10.1016/j.canlet.2017.04.028
82. Sharma SV et al (2010) A chromatin-mediated reversible drug-tolerant state in cancer cell subpopulations. Cell 141:69–80. https://doi.org/10.1016/j.cell.2010.02.027
83. Davies C et al (2013) Long-term effects of continuing adjuvant tamoxifen to 10 years versus stopping at 5 years after diagnosis of oestrogen receptor-positive breast cancer: ATLAS, a randomised trial. Lancet 381:805–816. https://doi.org/10.1016/S0140-6736(12)61963-1
84. Liu XF, Bagchi MK (2004) Recruitment of distinct chromatin-modifying complexes by tamoxifen-complexed estrogen receptor at natural target gene promoters in vivo. J Biol Chem 279:15050–15058. https://doi.org/10.1074/jbc.M311932200
85. Stone A et al (2012) Tamoxifen-induced epigenetic silencing of oestrogen-regulated genes in anti-hormone resistant breast cancer. PLoS ONE 7:e40466. https://doi.org/10.1371/journal.pone.0040466
86. Jimenez-Garduno AM et al (2017) IL-1beta induced methylation of the estrogen receptor ERalpha gene correlates with EMT and chemoresistance in breast cancer cells. Biochem Biophys Res Commun 490:780–785. https://doi.org/10.1016/j.bbrc.2017.06.117
87. Fang M, Hutchinson L, Deng A, Green MR (2016) Common BRAF(V600E)-directed pathway mediates widespread epigenetic silencing in colorectal cancer and melanoma. Proc Natl Acad Sci U S A 113:1250–1255. https://doi.org/10.1073/pnas.1525619113
88. Hou P, Liu D, Dong J, Xing M (2012) The BRAF(V600E) causes widespread alterations in gene methylation in the genome of melanoma cells. Cell Cycle 11:286–295. https://doi.org/10.4161/cc.11.2.18707
89. Zakharia Y et al (2017) Targeting epigenetics for treatment of BRAF mutated metastatic melanoma with decitabine in combination with vemurafenib: A phase lb study. Oncotarget 8:89182–89193. https://doi.org/10.18632/oncotarget.21269

90. Paoluzzi L et al (2016) BET and BRAF inhibitors act synergistically against BRAF-mutant melanoma. Cancer Med 5:1183–1193. https://doi.org/10.1002/cam4.667
91. Hurley PJ et al (2015) Androgen-regulated SPARCL1 in the tumor microenvironment inhibits metastatic progression. Cancer Res 75:4322–4334. https://doi.org/10.1158/0008-5472.CAN-15-0024
92. Thompson VC et al (2012) A gene signature identified using a mouse model of androgen receptor-dependent prostate cancer predicts biochemical relapse in human disease. Int J Cancer 131:662–672. https://doi.org/10.1002/ijc.26414
93. Karantanos T, Corn PG, Thompson TC (2013) Prostate cancer progression after androgen deprivation therapy: mechanisms of castrate resistance and novel therapeutic approaches. Oncogene 32:5501–5511. https://doi.org/10.1038/onc.2013.206
94. Gaspar-Maia A, Alajem A, Meshorer E, Ramalho-Santos M (2011) Open chromatin in pluripotency and reprogramming. Nat Rev Mol Cell Biol 12:36–47. https://doi.org/10.1038/nrm3036
95. Yamanaka S, Takahashi K (2006) Induction of pluripotent stem cells from mouse fibroblast cultures. Tanpakushitsu Kakusan Koso 51:2346–2351
96. Kim J et al (2010) A Myc network accounts for similarities between embryonic stem and cancer cell transcription programs. Cell 143:313–324. https://doi.org/10.1016/j.cell.2010.09.010
97. Kim J, Chu J, Shen X, Wang J, Orkin SH (2008) An extended transcriptional network for pluripotency of embryonic stem cells. Cell 132:1049–1061. https://doi.org/10.1016/j.cell.2008.02.039
98. Scognamiglio R et al (2016) Myc depletion induces a pluripotent dormant state mimicking diapause. Cell 164:668–680. https://doi.org/10.1016/j.cell.2015.12.033
99. Liu Z et al (2012) Regulation of p27Kip1 by Sox2 maintains quiescence of inner pillar cells in the murine auditory sensory epithelium. J Neurosci 32:10530–10540. https://doi.org/10.1523/JNEUROSCI.0686-12.2012
100. Surzenko N, Crowl T, Bachleda A, Langer L, Pevny L (2013) SOX2 maintains the quiescent progenitor cell state of postnatal retinal Muller glia. Development 140:1445–1456. https://doi.org/10.1242/dev.071878
101. Avincsal MO et al (2017) Epigenetic down-regulation of SOX2 is an independent poor prognostic factor for hypopharyngeal cancers. Histopathology. https://doi.org/10.1111/his.13436
102. Cheung TH, Rando TA (2013) Molecular regulation of stem cell quiescence. Nat Rev Mol Cell Biol 14:329–340. https://doi.org/10.1038/nrm3591
103. Avgustinova A, Benitah SA (2016) Epigenetic control of adult stem cell function. Nat Rev Mol Cell Biol 17:643–658. https://doi.org/10.1038/nrm.2016.76
104. Boonsanay V et al (2016) Regulation of skeletal muscle stem cell quiescence by Suv4-20h1-dependent facultative heterochromatin formation. Cell Stem Cell 18:229–242. https://doi.org/10.1016/j.stem.2015.11.002
105. Cabezas-Wallscheid N et al (2017) Vitamin A-retinoic acid signaling regulates hematopoietic stem cell dormancy. Cell 169:807–823 e819. https://doi.org/10.1016/j.cell.2017.04.018
106. Ghajar CM (2015) Metastasis prevention by targeting the dormant niche. Nat Rev Cancer 15:238–247. https://doi.org/10.1038/nrc3910
107. Zhang X, Cruz FD, Terry M, Remotti F Matushansky I (2013) Terminal differentiation and loss of tumorigenicity of human cancers via pluripotency-based reprogramming. Oncogene 32, 2249–2260, 2260 e2241–2221, doi:10.1038/onc.2012.237
108. Kollermann MW et al (1998) Supersensitive PSA-monitored neoadjuvant hormone treatment of clinically localized prostate cancer: effects on positive margins, tumor detection and epithelial cells in bone marrow. Eur Urol 34:318–324
109. Becker S, Solomayer E, Becker-Pergola G, Wallwiener D, Fehm T (2007) Primary systemic therapy does not eradicate disseminated tumor cells in breast cancer patients. Breast Cancer Res Treat 106:239–243. https://doi.org/10.1007/s10549-006-9484-5
110. Becker S, Becker-Pergola G, Wallwiener D, Solomayer EF, Fehm T (2006) Detection of cytokeratin-positive cells in the bone marrow of breast cancer patients undergoing adjuvant therapy. Breast Cancer Res Treat 97:91–96. https://doi.org/10.1007/s10549-005-9095-6
111. Naume B et al (2014) Clinical outcome with correlation to disseminated tumor cell (DTC) status after DTC-guided secondary adjuvant treatment with docetaxel in early breast cancer. J Clin Oncol 32:3848–3857. https://doi.org/10.1200/JCO.2014.56.9327
112. Solomayer EF et al (2012) Influence of zoledronic acid on disseminated tumor cells in primary breast cancer patients. Ann Oncol 23:2271–2277. https://doi.org/10.1093/annonc/mdr612
113. Banys M et al (2013) Influence of zoledronic acid on disseminated tumor cells in bone marrow and survival: results of a prospective clinical trial. BMC Cancer 13:480. https://doi.org/10.1186/1471-2407-13-480
114. Ashworth TR (1869) A case of cancer in which cells similar to those in the tumours were seen in the blood after death. Aus Med J 14:146–147
115. Zhang L et al (2012) Meta-analysis of the prognostic value of circulating tumor cells in breast cancer. Clin Cancer Res Off J Am Assoc Cancer Res 18:5701–5710. https://doi.org/10.1158/1078-0432.CCR-12-1587
116. Hayes DF et al (2006) Circulating tumor cells at each follow-up time point during therapy of metastatic breast cancer patients predict progression-free and overall survival. Clin Cancer Res Off J Am Assoc Cancer Res 12:4218–4224. https://doi.org/10.1158/1078-0432.CCR-05-2821

117. Maltoni R, Gallerani G, Fici P, Rocca A, Fabbri F (2016) CTCs in early breast cancer: a path worth taking. Cancer Lett 376:205–210. https://doi.org/10.1016/j.canlet.2016.03.051
118. Fischer JC et al (2013) Diagnostic leukapheresis enables reliable detection of circulating tumor cells of nonmetastatic cancer patients. Proc Natl Acad Sci U S A 110:16580–16585. https://doi.org/10.1073/pnas.1313594110
119. Fehm T et al (2009) Detection and characterization of circulating tumor cells in blood of primary breast cancer patients by RT-PCR and comparison to status of bone marrow disseminated cells. Breast Cancer Res 11:R59. https://doi.org/10.1186/bcr2349
120. Molloy TJ et al (2011) The prognostic significance of tumour cell detection in the peripheral blood versus the bone marrow in 733 early-stage breast cancer patients. Breast Cancer Res 13:R61. https://doi.org/10.1186/bcr2898
121. Benoy IH et al (2006) Real-time RT-PCR detection of disseminated tumour cells in bone marrow has superior prognostic significance in comparison with circulating tumour cells in patients with breast cancer. Br J Cancer 94:672–680. https://doi.org/10.1038/sj.bjc.6602985
122. Wiedswang G et al (2006) Comparison of the clinical significance of occult tumor cells in blood and bone marrow in breast cancer. Int J Cancer 118:2013–2019. https://doi.org/10.1002/ijc.21576
123. Shaw JA et al (2012) Genomic analysis of circulating cell-free DNA infers breast cancer dormancy. Genome Res 22:220–231. https://doi.org/10.1101/gr.123497.111
124. Widschwendter M et al (2017) Methylation patterns in serum DNA for early identification of disseminated breast cancer. Genome Med 9:115. https://doi.org/10.1186/s13073-017-0499-9
125. Sigalotti L et al (2012) Whole genome methylation profiles as independent markers of survival in stage IIIC melanoma patients. J Transl Med 10:185. https://doi.org/10.1186/1479-5876-10-185
126. Jiang W et al (2015) Genome-wide identification of a methylation gene panel as a prognostic biomarker in nasopharyngeal carcinoma. Mol Cancer Ther 14:2864–2873. https://doi.org/10.1158/1535-7163.MCT-15-0260
127. Akhtar-Zaidi B et al (2012) Epigenomic enhancer profiling defines a signature of colon cancer. Science 336:736–739. https://doi.org/10.1126/science.1217277
128. Lian CG et al (2012) Loss of 5-hydroxymethylcytosine is an epigenetic hallmark of melanoma. Cell 150:1135–1146. https://doi.org/10.1016/j.cell.2012.07.033
129. Johnson KC et al (2016) 5-Hydroxymethylcytosine localizes to enhancer elements and is associated with survival in glioblastoma patients. Nat Commun 7:13177. https://doi.org/10.1038/ncomms13177
130. Kafer GR et al (2016) 5-hydroxymethylcytosine marks sites of DNA damage and promotes genome stability. Cell Rep 14:1283–1292. https://doi.org/10.1016/j.celrep.2016.01.035
131. Fernandez AF et al (2012) A DNA methylation fingerprint of 1628 human samples. Genome Res 22:407–419. https://doi.org/10.1101/gr.119867.110
132. Mattout A, Biran A, Meshorer E (2011) Global epigenetic changes during somatic cell reprogramming to iPS cells. J Mol Cell Biol 3:341–350. https://doi.org/10.1093/jmcb/mjr028
133. Theunissen TW et al (2011) Nanog overcomes reprogramming barriers and induces pluripotency in minimal conditions. Curr Biol 21:65–71. https://doi.org/10.1016/j.cub.2010.11.074
134. Lin B et al (2015) Global analysis of H3K4me3 and H3K27me3 profiles in glioblastoma stem cells and identification of SLC17A7 as a bivalent tumor suppressor gene. Oncotarget 6:5369–5381. https://doi.org/10.18632/oncotarget.3030
135. Trowbridge JJ et al (2012) Haploinsufficiency of Dnmt1 impairs leukemia stem cell function through derepression of bivalent chromatin domains. Genes Dev 26:344–349. https://doi.org/10.1101/gad.184341.111
136. Yang CH et al (2014) Identification of CD24 as a cancer stem cell marker in human nasopharyngeal carcinoma. PLoS ONE 9:e99412. https://doi.org/10.1371/journal.pone.0099412
137. Tung PY, Knoepfler PS (2015) Epigenetic mechanisms of tumorigenicity manifesting in stem cells. Oncogene 34:2288–2296. https://doi.org/10.1038/onc.2014.172
138. Price TT et al (2016) Dormant breast cancer micrometastases reside in specific bone marrow niches that regulate their transit to and from bone. Sci Transl Med 8:340ra373. https://doi.org/10.1126/scitranslmed.aad4059

2 Anti-angiogenic Therapy-Mediated Endothelial Damage: A Driver of Breast Cancer Recurrence?

Laura Pisarsky and Cyrus M. Ghajar

Abstract

Anti-angiogenic therapy was conceived originally as a silver bullet able to maintain tumor dormancy indefinitely. By targeting new blood vessel formation, anti-angiogenic agents were expected to suppress the growth of any type of primary or metastatic tumor, independent of their subtype or genetic landscape. However, more that 20 years after the first anti-angiogenic preclinical trial, the astonishing inhibition of metastatic outgrowth originally observed in mouse models never translated into clinics. Indeed, whereas anti-angiogenic agents (sometimes) prolong progression-free survival, they fail to impact overall survival, particularly in breast cancer. This observation revealed to be true in early- and advanced-stage breast cancer patients treated either in adjuvant or neo-adjuvant settings, suggesting that the effect of anti-angiogenic therapy on repressing growth of overt metastases – and also on *preventing* outgrowth of disseminated tumor cells and micrometastases – is limited. What are the reasons underlying this failure? And, more importantly, is there still room for improvement?

Keywords

Anti-angiogenic therapy · Breast cancer · Endothelial damage · Tumor dormancy · Micrometastases · DTC · CTCs · Angiogenesis · Intra- and extra-vasation · Perivascular niches · MRD

Abbreviations

ARP-2/3	Actin-related protein-2/3
BMP	Bone morphogenic protein
CCL-2	CC-chemokine ligand-2
CCR-2	CC-chemokine receptor-2
CSC	Cancer stem cell
CTC	Circulating tumor cell
CXCL-12	CXC-chemokine ligand-12
CXCR-4	CXC-chemokine receptor-4
Dll4	Delta-like protein 4
DTC	Disseminated tumor cell
EMT	Epithelial-mesenchymal transition
ERK	Extracellular signal-regulated kinase
FAK	Focal adhesion kinase
GAS6	Growth arrest-specific gene 6

L. Pisarsky (✉) · C. M. Ghajar (✉)
Translational Research Program, Public Health Sciences Division, Fred Hutchinson Cancer Research Center, Seattle, WA, USA
e-mail: lpisarsk@fredhutch.org; cghajar@fredhutch.org

J. A. Aguirre-Ghiso (ed.), *Biological Mechanisms of Minimal Residual Disease and Systemic Cancer*, Advances in Experimental Medicine and Biology 1100,
https://doi.org/10.1007/978-3-319-97746-1_2

G-CSF	Granulocyte-colony stimulating factor
HGF	Hepatocytes growth factor
IGF-1	Insulin-like growth factor-1
IGF-1R	Insulin-like growth factor-1 receptor
IGFBP-7	Insulin-like growth factor binding protein-7
MMP-14	Matrix metalloproteinase-14
PDGFR	Platelet-derived growth factor receptor
PVN	Perivascular niche
SCF	Stem cell factor
SDF-1	Stromal cell-derived factor-1
TGF-β	Transforming growth factor-β
uPAR	Urokinase-type plasminogen activator receptor
VCAM-1	Vascular cell adhesion molecule-1
VEGF	Vascular endothelial growth factor
VEGFR	Vascular endothelial growth factor receptor

2.1 Introduction

Thirty percent of breast cancer survivors still harbor circulating tumor cells (CTCs) years or decades after mastectomy, and indeed, breast cancers often recur even after such a long latency [95, 150]. Undetectable disseminated tumor cells (DTCs) have been suggested to feed the CTC pool [100]. In breast cancer patients, presence of DTCs in the bone marrow at diagnosis is associated with a higher risk of metastatic recurrence not only in the bone marrow, but across all metastatic sites [17, 109, 157]. Association between DTCs in the bone marrow and poor clinical outcome has been observed in several solid cancers and interestingly, even in cancer types that rarely form overt bone metastases, such as esophageal cancer [146]. Importantly, effective elimination of such residual tumor cells by additional cycles of chemotherapy significantly improves disease-free survival in early breast cancer patients, implicating these cells as the source of late recurrence [9, 109]. However, there is, to date, no therapy able to completely eradicate dormant DTCs or to prevent their reactivation and growth as overt metastases.

Angiogenesis, the formation of new blood vessels from a pre-existing vasculature, is crucial to several physiological and pathological processes, such as organ development and cancer [26]. In order to sustain a high proliferation rate, developing tumors require access to nutrients and oxygen and therefore rely on the efficient recruitment of blood vessels [73]. Indeed, dating back to 1939, the work of Ide and collaborators suggested that tumors implanted in rabbit ears produce a vessel-growth stimulating factor, observations that were confirmed by others in additional animal models [5, 50, 69, 77]. Later, fragments of rabbit carcinomas implanted into the anterior chamber of the eyes of guinea pigs were shown to enter a phase of latency until they received visible blood supply [70], a phenomenon today known as the "angiogenic switch [72]". Based on these data, Folkman proposed that inhibition of blood supply might maintain tumor cells in a dormant state, thereby preventing metastasis [56]. The purification in 1989 of the main regulator of angiogenesis – vascular endothelial growth factor (VEGF) – led to the development of a large array of anti-angiogenic therapies [55, 89]. The first of them, a neutralizing antibody targeting VEGF-A named bevacizumab, was approved as single agent therapy for the treatment of recurrent glioblastoma, and in combination with chemotherapy in metastatic colorectal cancer, non-small cell lung carcinoma and metastatic renal cell cancer [32, 53, 81]. Based on its ability to extend progression-free survival of metastatic breast cancer patients in the E2100 trial, bevacizumab benefited from accelerated FDA approval in combination with paclitaxel for this condition [104] (Table 2.1). However, absence of prolonged overall survival of the patients, together with more moderated effect on their progression-free survival in the RIBBON-1 and AVADO trials, ultimately led to its withdrawal 2 years later in the USA [119, 130].

Indeed, resistance to bevacizumab rapidly occurs, driven at least in part by the redundancy of angiogenic molecules [14, 128]. To compensate for the activation of alternative pro-angiogenic

Table 2.1 Randomized clinical trials assessing the efficacy of anti-angiogenics in addition to or in place of chemotherapy or endocrinotherapy in breast cancer patients

	Trial	Patients	Treatment settings	Tumor type and stage	Therapeutic arms	PFS	OS
Bevacizumab	NCT01250379 TANIA [152, 153]	494	Bevacizumab after progression on first-line bevacizumab therapy	HER2-negative locally recurrent/ metastatic breast cancer	Chemotherapy + bevacizumab vs Chemotherapy alone	2nd line: 3.8 vs 2.9 months (P = 0.1080) 3rd line: 12.8 vs 10.7 months (P = 0.1349)	19.7 vs 18.7 months (P = 0.7253)
	NCT00028990 E2100 [104]	722	Paclitaxel plus bevacizumab versus paclitaxel alone	Metastatic breast cancer	Paclitaxel + bevacizumab vs Paclitaxel alone	11.8 vs 5.9 months (P ≤ 0.001)	26.7 vs 25.2 months (P = 0.16)
	NCT00281697 RIBBON-2 [20, 75]	684	Bevacizumab in combination with chemotherapy in second-line treatment after progression on chemotherapy	Human epidermal growth factor receptor 2-negative metastatic breast cancer	Chemotherapy + bevacizumab vs Chemotherapy + placebo	7.2 vs 5.1 months (P = 0.0072)	18.6 vs 17.8 months (P = 0.8843)
	NCT00262067 RIBBON-1 [129, 130]	1237	Chemotherapy with or without bevacizumab as first-line treatment	Human epidermal growth factor receptor 2-negative, locally recurrent or metastatic breast cancer	Capecitabine + bevacizumab vs Capecitabine + placebo	8.6 vs 5.7 months (P ≤ 0.001)	1-year survival rate: 81.0% vs 74.4% (P = 0.076)
					Taxane/anthracycline + bevacizumab vs Taxane/ anthracycline + placebo	9.2 vs 8.0 months (P ≤ 0.001)	80.7% vs 83.2% (P = 0.44).
	NCT00391092 AVEREL [66]	424	Bevacizumab in combination with trastuzumab and docetaxel as first-line treatment	HER2-positive locally recurrent/ metastatic breast cancer	Trastuzumab + docetaxel + bevacuzumab vs Trastuzumab + docetaxel	16.5 vs 13.7 months (P = 0.0775)	38.5 vs 38.3 months (P = 0.9543)
	NCT01663727 MERiDiAN [101, 102]	481	Bevacizumab plus paclitaxel versus placebo plus paclitaxel as first line treatment	HER2-negative metastatic breast cancer	Paclitaxel + bevacizumab vs Paclitaxel + placebo	11.0 vs 8.8 months (P = 0.0007)	28.8 vs 25.8 months (P = 0.59)

(continued)

Table 2.1 (continued)

	Trial	Patients	Treatment settings	Tumor type and stage	Therapeutic arms	PFS	OS
	NCT00856492 SWOG S0800 [108]	211	Chemotherapy with or without bevacizumab	Inflammatory or locally advanced breast cancer	Chemotherapy + bevavizumab vs Chemotherapy alone	N/D	3-year overall survival: 86 vs 87% (P = 0.64)
	NCT00408408 [10, 11]	1206	Neoadjuvant chemotherapy with or without bevacizumab	Stage I, stage II, or stage IIIA breast cancer	Chemotherapy + bevacizumab vs Chemotherapy alone	N/D Rate of pathological complete response in the breast and lymph nodes: 27.6% vs 23.0% (P = 0.08)	N/D
	NCT00520975[a]	96	Chemotherapy and trastuzumab with or without bevacizumab as first-line treatment	Metastatic breast cancer overexpressing HER2/NEU	Chemotherapy + bevacizumab vs Chemotherapy + placebo	13.8 vs 11.1 months (P = 0.24)	63.0 vs 49.1 months (P = 0.75)
	NCT00601900 CALGB 4050 [40]	391	Tamoxifen citrate or letrozole with or without bevacizumab as first-line treatment	Stage III or stage IV breast cancer	Endocrine therapy + bevacizumab vs Endocrine therapy alone	20.2 vs 15.6 months (P = 0.016)	47.2 vs 43.9 months (P = 0.19)
	NCT00528567 BEATRICE [13, 23]	2591	Bevacizumab adjuvant therapy	Triple negative breast cancer	Chemotherapy + bevacizumab vs Chemotherapy alone	5-year invasive-disease-free survival rate: 79.6 vs 76.9% (no statistical analysis provided)	5-year overall survival rate: 87.9 vs 87.7% (P = 0.5247)
	NCT0333775 [103]	736	Docetaxel with or without bevacizumab as first-line treatment	HER2-negative metastatic breast cancer	Docetaxel + bevacizumab 15 mg/kg vs Docetaxel + bevacizumab 7.5 mg/kg vs Docetaxel + placebo	10.0 vs 9.0 vs 8.1 months ($P \leq 0.001$)	N/D

	NCT00356681 [97]	282	Motesanib, or open-label bevacizumab, in combination with paclitaxel as first-line treatment	HER2-negative locally recurrent or metastatic breast cancer	Paclitaxel + bevacizumab vs Paclitaxel + motesanib vs Paclitaxel + placebo	11.5 (P = 0.24) vs 9.5 (P = 0.79) vs 9.0 months	N/D
	NCT00545077 [96]	380	Addition of bevacizumab to endocrine therapy as first-line treatment	Advanced breast cancer	Letrozole/fulvestrant + bevacizumab vs Letrozole/fulvestrant	19.3 vs 14.4 months (P = 0.126)	52.1 vs 51.8 months (P = 0.518)
	NCT01093235 ARTemis [45]	800	Neoadjuvant bevacizumab added to docetaxel followed by fluorouracil, epirubicin and cyclophosphamide	HER2-negative early breast cancer	Docetaxel + bevacizumab -> fluorouracil + epirubicin + cyclophosphamide vs Docetaxel -> fluorouracil + epirubicin + cyclophosphamide	Disease-free survival events: 26 vs 22% (P = 0.25)	Overall survival events: 19 vs 15% (P = 0.19)
	NCT00567554 GBG44-GeparQuinto [151]	1948	Neoadjuvant chemotherapy with or without bevacizumab or everolimus	HER2-negative primary breast cancer	Epirubicine/cyclophosphamide + bevacizumab vs Epirubicine/cyclophosphamide	Estimated 3-year disease-free survival: 80.0 vs 81.5% (P = 0.784)	Estimated 3-year overall survival: 90.7 vs 88.7% (P = 0.842)
	NCT00511459 [41]	228	Trebananib plus weekly paclitaxel with or without bevacizumab as first-line therapy	HER2-negative locally recurrent or metastatic breast cancer	Paclitaxel + bevacizumab + trebananib 10 mg/kg once weekly vs Paclitaxel + bevacizumab + trebananib 3 mg/kg once weekly vs Paclitaxel + bevacizumab + placebo	11.3 (P = 0.946) vs 9.2 (P = 0.642) vs 12.2 months	N/D
Sunitinib	NCT00246571 [37]	217	Sunitinib versus standard of care in patients relapsing after anthracycline and taxane treatment	Advanced triple negative breast cancer	Sunitinib vs Standard of care chemotherapy	2.0 vs 2.7 months (P = 0.888)	9.4 vs 10.5 months (P = 0.839)

(continued)

Table 2.1 (continued)

	Trial	Patients	Treatment settings	Tumor type and stage	Therapeutic arms	PFS	OS
	NCT00393939 [15]	593	Sunitinib in combination with docetaxel versus docetaxel alone as first-line treatment	Advanced breast cancer	Docetaxel + sunitinib vs Docetaxel alone	8.6 vs 8.3 months (P = 0.265)	24.8 vs 25.5 months (P = 0.904)
	NCT00435409 [36]	442	Sunitinib in combination with capecitabine versus capecitabine monotherapy in patients previously treated with anthracycline and a taxane	Metastatic breast cancer	Capecitabine + sunitinib vs Capecitabine	5.5 vs 5.9 months (P = 0.941)	16.4 vs 16.5 months (P = 0.484)
	NCT00373113[a]	482	Sunitinib vs capecitabine after progression on taxane alone or taxane and anthracycline	Breast cancer	Sunitinib vs capecitabine	2.8 vs 4.2 months (P = 0.002)	15.3 vs 16.9 months (P = 0.219)
	NCT00373256 [131]	485	Sunitinib plus paclitaxel versus bevacizumab plus paclitaxel as first-line treatment	Advanced breast cancer	Paclitaxel + sunitinib vs Paclitaxel + bevacizumab	7.4 vs 9.2 months (P = 0.9986)	2-year survival rate: 36 vs 61% (no statistical analysis provided)
	NCT01803503 [8]	20	Docetaxel with or without low-dose, short course sunitinib	Advanced breast cancer	Docetaxel + sunitinib vs Docetaxel alone	11.4 vs 5.3 months (P = 0.101)	N/D
	NCT01176799 [158]	49	Doxorubicin and cyclophosphamide with or without intermittent sunitinib in the first-line treatment	Locally advanced or metastatic breast cancer with measurable primary breast tumor	Chemotherapy + sunitinib vs Chemotherapy alone	Relapse-free survival: 37.0 vs 37.1 months (P = 0.70)	47.6 vs 49.7 months (P = 0.74)

Sorafenib	NCT00493636 [134]	121	Sorafenib or placebo with either gemcitabine or capecitabine in patients that progressed during or after bevacizumab	HER2-negative advanced breast cancer	Gembcitabine/capecitabine + sorafenib vs Gembcitabine/capecitabine + placebo	3.4 vs 2.7 months (P = 0.02)	13.4 vs 11.4 months (P = 0.95)

PFS Progression-free survival, *OS* Overall survival, *N/D* Non-defined
[a]Based on the results published on www.clinicaltrials.org

signaling pathways, tyrosine kinase inhibitors with a broad range of inhibition were developed. Of these, sunitinib is now approved for renal cell carcinoma and pancreatic neuroendocrine tumors, sorafenib is the standard of care in the treatment of hepatocellular and kidney cancers, and many additional anti-angiogenic strategies are currently under evaluation [52, 90, 126, 127]. However, their use in breast cancer have, once more, ended in failure. For example, sunitinib was associated at best with lack of improvement and at worst with significantly decreased progression-free survival (Table 2.1). These unexpected results obtained with both first- and second-generation anti-angiogenics have left us with quite a puzzle to piece together.

In this chapter, we will discuss possible reasons underlying this lack of efficacy, focusing on how anti-angiogenic-induced endothelial damage may promote metastases. We will describe how anti-angiogenic therapies can increase tumor cell invasiveness and dissemination from primary tumors (Sect. 2.2), and subsequently discuss the potential effects of anti-angiogenic therapies on metastatic colonization (Sect. 2.3), a topic that remains poorly studied.

2.2 Anti-angiogenics in the Early Steps of Metastasis

2.2.1 Blood Vessels as Central Actors of Metastatic Dissemination

Dissemination of cancer cells via blood or lymphatic routes, followed by efficient colonization of distant organs represent the main cause of breast cancer-related death in patients. The vasculature represents a major actor in cancer progression, providing nutrients to the primary tumor and guiding DTCs to their next destination. Moreover, once they reach distant organs, tumor cells bind to the endothelial wall, extravasate and settle on the basal side of the endothelium, where endothelial cell-derived factors dictate their ability to survive, and regulate DTC dormancy and proliferation [63]. Establishment of neovasculature at distant sites finally allows metastatic outgrowth.

Usage of anti-angiogenic therapy in neoadjuvant settings – preceding primary tumor resection – was expected to starve the primary tumor to death, resulting in its shrinkage, thus reducing the size of surgical margins. By decreasing the number of intratumoral blood vessels, one could imagine that tumor cell entrance into the bloodstream would be limited, therefore preventing cancer dissemination. In adjuvant settings – following primary tumor resection – anti-angiogenics aimed to maintain DTCs and metastases indolent by thwarting recruitment of new blood vessels. Unfortunately, clinical reality fell far short of the expectations. Effects are limited and some tumor types, such as breast cancer, are particularly resistant to angiogenesis inhibition (Table 2.1). Yet, we have not been able to fully understand the mechanisms underlying resistance to anti-angiogenics, let alone overcome them. Is the dearth of efficacy of anti-angiogenic therapies simply due to a lack of reliance of some tumor types on angiogenesis? Or, are we facing a more complex problem?

2.2.2 Anti-angiogenic Therapy Fosters Metastatic Dissemination

Several studies have described a significant reduction of primary tumor growth, invasion and metastasis following anti-angiogenic therapy [30, 61]. However, others reported increased tumor cell migration and accelerated metastatic disease. Indeed, Ebos and colleagues conducted a pivotal study, observing that despite beneficial effects on primary tumors, short-term neoadjuvant sunitinib treatment fosters metastasis development [48]. This lack of correlation between the response of primary tumors and metastatic lesions to anti-angiogenic therapy has been further validated in additional cancer models [31, 49, 71]. Data from these studies help explain the unsuccessful translation of anti-angiogenics from laboratories – where they were essentially tested against primary tumors –

to clinics – where they are principally used in metastatic patients (Table 2.1).

Unlike sunitinib treatment, extracellular anti-angiogenic therapies such as B-20 – a VEGF targeting antibody – or DC-101 – an antibody able to bind the ligand-binding site of VEGFR-2 – did not increase metastases in breast cancer and melanoma models [31, 49]. The fact that "mild" anti-angiogenic treatments, such as B-20, did not accelerate metastases, whereas broad spectrum anti-angiogenic agents fostered metastatic outgrowth [49], suggests that active disruption of vascular homeostasis might be more damaging than any benefit derived from inhibition of vascular sprouting.

Interestingly, however, when administered for several weeks, DC-101 treatment increased invasion in primary MDA-MB-231 tumors and breast cancer patient-derived xenograft models, an effect that could be reverted by combining DC-101 with paclitaxel or cyclophosphamide [111]. Moreover, combination treatment impaired tumor cell seeding and outgrowth within the lung. In a similar manner, neo-adjuvant treatment of mice bearing mammary tumors using a DC-101/paclitaxel combination significantly improved overall survival [111].

Altogether, these data suggest that anti-angiogenic therapy increases metastatic burden, and that this effect can be reverted by combining anti-angiogenics with chemotherapy. What are the molecular mechanisms leading to increased metastatic spread following anti-angiogenic therapy? Data accumulated in recent years suggest that damaging the endothelium, which is essentially what these agents do, affects virtually every step of the metastatic cascade.

2.2.3 Anti-angiogenic Therapy Enriches for a Highly Metastatic Population

Hungry lions like to eat, and starving tumors is not without consequences. Whereas their initial response is to shrink, establishment of a hypoxic program rapidly results in tumor revascularization through alternative pro-angiogenic signaling pathways activated by the tumor cells themselves or via the recruitment of bone marrow-derived cells [14, 44, 138]. When necessary, tumor cells cope with hypoxia by adapting their metabolism to survive in an oxygen-deprived environment [7, 80, 117]. Alternately, tumor cells might gain invasive ability, allowing them to escape the hypoxic and nutrient-deprived area [14, 82] (Fig. 2.1a, b).

Anti-angiogenic therapy-induced hypoxia has been associated with increased primary tumor invasiveness, particularly in models of insulinoma and glioblastoma. For example, in RIP1-Tag2 mice treated with DC-101, the incidence of highly invasive carcinoma is 10-fold higher than those observed in untreated animals [110]. Furthermore, in DC-101-treated animals, tumor cells did not only invade the local surrounding microenvironment but also the vascular space, resulting in 2.7-times more metastatic lesions. Paez-Ribes and colleagues validated these findings using sunitinib, but also by employing a glioblastoma model. In agreement with these data, renal cell carcinoma patients exhibit a transient increase in CTCs following their first cycle of sunitinib treatment [106].

Epithelial-mesenchymal transition (EMT) and induction of a stem-like phenotype have been implicated in hypoxia-induced invasiveness. For instance, transcription of the proto-oncogene Met is induced under hypoxic pressure, and is associated with EMT [35, 114]. Alternatively, in pancreatic ductal adenocarcinoma, hypoxia induced transforming growth factor (TGF)-β, resulting in EMT upon VEGF blockade [1]. Hypoxia and EMT have been associated with cancer stem cell (CSC) enrichment, an indicator of poor prognosis in cancer patients [54, 94]. Hence, by fostering hypoxia and an invasive phenotype, anti-angiogenic treatment might increase tumor stemness potential. Indeed, in breast cancer animal models, sunitinib-induced hypoxia enriches for CSCs and is, therefore, expected to increase metastatic colonization [33, 34].

Altogether, these data suggest that hypoxia-associated metastatic dissemination can be reverted by reoxygenation of the tumor or inhibition of hypoxic response genes [28, 99, 136].

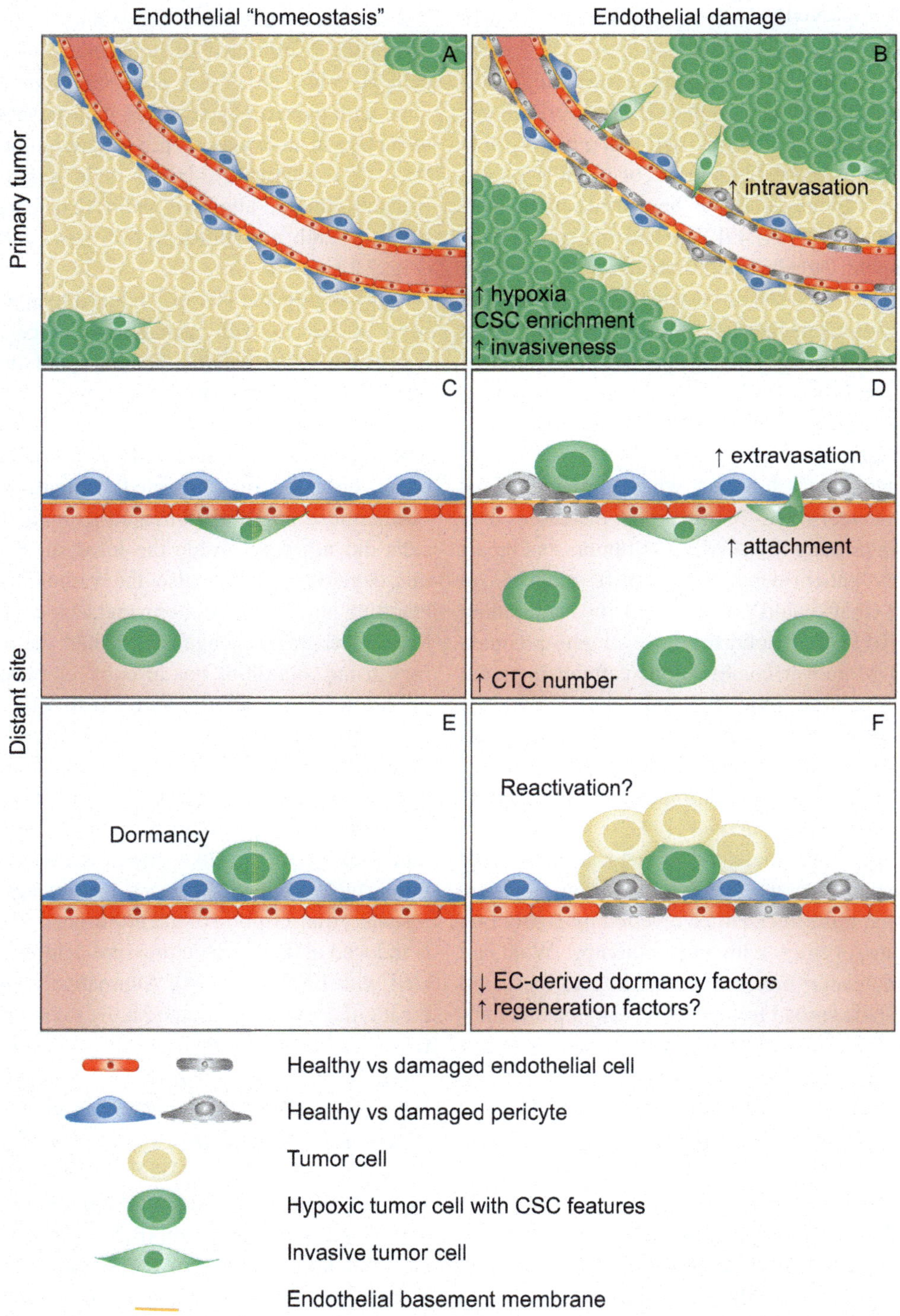

Fig. 2.1 Effects of endothelial damage on metastatic progression. Anti-angiogenic therapy induces endothelial cell and pericyte apoptosis, leading to increased hypoxia in primary tumors (**a–b**). As a result, tumor cells gain an invasive phenotype and their access to the circulation is further facilitated by vascular leakiness. Consequently, an increased number of CTCs reaches metastatic sites, where denuded basement membrane facilitates their binding and extravasation (**c–d**). Finally, we hypothesize that damaged endothelium in distant organs is associated with a loss of endothelial cell-derived dormancy factors and increased secretion of inflammatory cytokines and regeneration-mediating growth factors, which could reactivate DTC proliferation (**e–f**)

2.2.4 Vascular Damage and Tumor Cell Intra- and Extra-Vasation

The aforementioned examples provide strong evidence that anti-angiogenic therapy creates a harsh environment at primary site, in which only the fittest tumor cells will survive. However, those examples are not sufficient to explain the increased metastatic burden following anti-angiogenic treatment. Indeed, some studies have suggested that anti-angiogenic therapy accelerates metastatic outgrowth even in the absence of selective pressure at the primary site [48]. For example, Ebos and colleagues demonstrated that intravenous tumor cell injection leads to increased metastatic burden in mice pre-treated with sunitinib compared to vehicle-treated mice. These data suggest that the damage caused by anti-angiogenics at distant sites could facilitate either the recruitment, extravasation and/or outgrowth of DTCs (Fig. 2.1c, d).

There is, to date, very limited data regarding the response of stable vessels to anti-angiogenic therapy. However, whereas anti-angiogenics have been designed to inhibit the formation of new blood vessels – as opposed to vascular disrupting agents targeting both neo-angiogenesis and already established and mature vessels, it is clear that they also damage quiescent vessels and may damage mature endothelium at metastatic sites.

In cancer patients, loss of vascular integrity following anti-angiogenic regimens is supported by the bleeding complications resulting from these treatments [51]. As described by Miller and colleagues, 28.8% of breast cancer patients treated with a bevacizumab/capecitabine combination experience bleedings, versus 11.2% in the placebo/capecitabine-treated arm [105]. Increased incidence of high blood pressure and thrombotic events in anti-angiogenic-treated patients also suggest endothelial injury [124]. Disruption of vascular homeostasis subsequent to anti-angiogenic treatment has also been observed in animal models. Indeed, Chung and collaborators observed tracheal vessel pruning in mice treated with anti-VEGF antibody or sunitinib [31]. Their work implicated vascular endothelial (VE)-cadherin repression in tumor cell extravasation. Cultured endothelial cells treated with sunitinib lost VE-cadherin phosphorylation, which resulted in alteration of adherens junctions and increased extravasation in animal models. These data are in line with the study of Weis and colleagues, which revealed that direct inhibition of VE-cadherin facilitates tumor cell extravasation [155].

Further evidence that anti-angiogenic therapies impact quiescent vessels comes from the observation that inhibition of intracellular VEGFR-2 signaling disrupts vascular homeostasis in mice [88]. Small tyrosine kinase inhibitors such as sunitinib penetrate the endothelial plasma membrane and inhibit autocrine VEGF-A signaling, leading to endothelial injury [88]. Moreover, by inhibiting platelet-derived growth factor receptors (PDGFRs), sunitinib also induces the detachment of pericytes from the abluminal endothelial membrane, fostering endothelial cell apoptosis [57]. These leaky vessels represent open doors that facilitate tumor cell intra- and extra-vasation [62, 159]. They also function to retain CTCs: following endothelial cell shedding, tumor cells bind "basement membrane ghosts [78]" via e.g., $\alpha 3\beta 1$ integrin, leading to their arrest at distant sites [154] (Fig. 2.1c, d).

Altogether, the studies discussed in Sect. 2.2 support the hypothesis that anti-angiogenic compounds cause vascular damage and promote tumor cell intravasation, arrest and extravasation. But even if tumor cells can disseminate at higher rates to distant sites, they still must colonize these sites to truly present a problem. Do anti-angiogenics aid or hinder this process?

2.3 Anti-angiogenics at Metastatic Sites

2.3.1 Primary Tumor, DTCs, Micro- and Macro-metastases: All Equal When It Comes to Angiogenesis?

Although anti-angiogenic therapy shows efficacy on mouse primary mammary tumors, seminal work from the Kerbel and Ferrara laboratories supports their lack of efficiency on metastatic disease [31, 46, 48, 71]. The data discussed in the first part of the chapter provide compelling evidence that anti-angiogenics can increase the release of tumor cells into the circulation and their extravasation at distant site. There is, however, very limited data regarding the effect of anti-angiogenics on metastatic colonization, and the factors implicated in the lack of response at metastatic sites remain elusive. Can we preclude that tumor cells, as they metastasize, undergo a sort of "angio-evolution"? In other terms, could an angiogenic primary tumor free itself from angiogenesis while becoming metastatic? Could the angiogenic switch, which is necessary for primary tumor outgrowth, become obsolete for DTCs exiting dormancy? Or could VEGF-dependent primary tumor cells rely simply on alternative angiogenic signaling pathways once at a metastatic site?

A study published by the Breast Cancer Progression Working Party is instructive in that regard [115]. Immunohistochemical analysis of primary breast tumors and lung metastases revealed that (i) primary tumors of patients presenting with lung metastases at follow-up were more vascularized than primary tumors of patients not presenting with metastases, supporting hematogeneous dissemination; (ii) lung metastases often were more vascularized than the primary tumors they originated from, suggesting a possible enrichment in angiogenic clones; (iii) a fifth of lung metastases developed in absence of neo-angiogenesis and might represent a population intrinsically resistant to anti-angiogenic agents; and (iv) both angiogenic and non-angiogenic metastases arose after a dormancy period of up to 6 years, challenging the concept of neo-angiogenesis as a rate-limiting factor in dormancy maintenance. These data have been supported by intravital imaging of the mouse brain [84]. Here, angiogenesis and vascular remodeling were not critical to the transition from single DTCs to micrometastases, but became essential to the transformation of micrometastases into macrometastases. Indeed, lung cancer micrometastases that were unable to initiate angiogenesis regressed, whereas melanoma cells relied on vessel co-option. In these settings, bevacizumab treatment impeded lung cancer-derived micrometastases from thriving. However, VEGF inhibition did not impact the outgrowth of melanoma cells, suggesting that vessel co-option might contribute to anti-angiogenic resistance.

Non-angiogenic metastatic outgrowth has been associated with vessel co-option, a mechanism in which tumor cells hijack vessels already present in their microenvironment. In breast cancer patients, non-angiogenic lung metastases invade alveolar air spaces and co-opt alveolar capillaries [18]. In colon cancer patients, half of the liver lesions in untreated patients are angiogenic, while the other half invade the surrounding microenvironment and co-opt nearby blood vessels [58]. Although this is also true for overt metastases treated with bevacizumab-chemotherapy, 80% of the lesions arising *after* bevacizumab-chemotherapy treatment are non-angiogenic, suggesting once again that dormant DTCs or micrometastases can resume growth without neo-angiogenesis, and that this is one mechanism by which DTCs overcome anti-angiogenic therapy. As a result, patients presenting with co-opting metastases do not benefit from anti-angiogenic addition and have an overall survival equivalent to patients treated with chemotherapy alone. Analysis of breast cancer liver metastases revealed that 94% of the analysed patients presented with sinusoidal vessel co-option, supporting the lack of efficacy of anti-angiogenics in this condition [58]. Finally, shRNA-mediated inhibition of actin-related protein (Arp)2/3 – a regulator of cancer cell motility – significantly reduced the colon cancer cell ability to co-opt blood vessels, and reduced metastatic outgrowth in mice when combined with anti-angiogenic therapy [58].

Most angiogenesis inhibitors currently used in clinics act through the inhibition of the VEGF signalling pathway. Nevertheless, not every tumor relies on VEGF-induced angiogenesis for their growth, resulting in intrinsic anti-angiogenic resistance. For instance, angiopoietin-targeting antibodies overcame intrinsic resistance to VEGF inhibition in adjuvant treatment of post-surgical breast and lung carcinomas [143]. Combination of anti-VEGF and anti-angiopoietin-2 antibodies with low dose metronomic gemcitabine maintained lung DTCs in an avascular, micrometastatic state. Here, angiopoeitein-2 inhibition is associated with a decreased secretion of CC-chemokine ligand (CCL)-2 by the lung endothelium. As a result, recruitment of CC-chemokine receptor (CCR)-2-positive metastasis-associated macrophages, known to be involved in anti-angiogenic resistance through their secretion of pro-angiogenic cytokines [122, 137, 138], was impaired [143]. Clinical trials are currently evaluating the efficacy of a bispecific angiopoietin-2-VEGF-A CrossMab in metastatic solid cancers [83]. In mouse models, angiopoietin-2-VEGF-A CrossMab significantly reduced the number of CTCs and the number of lung foci [83]. In post-surgical treatment settings, this CrossMab also inhibited metastatic outgrowth, suggesting that simultaneously targeting VEGF and angiopoitein-2 results both in decreased tumor cell shedding, but also in decreased lung metastatic burden [83].

Altogether, these studies support the lack of efficacy of VEGF-targeting angiogenesis inhibitors in metastatic breast cancer patients.

2.3.2 Anti-angiogenic Therapies in the Prevention of Disseminated Tumor Cell Outgrowth

In breast cancer patients, hematogeneous and lymphatic dissemination of cancer cells, followed by the efficient colonization of metastatic organs mark the transition from a local, resectable tumor to an incurable metastatic cancer. While this maiden voyage is initiated at the early steps of primary tumor formation, the emergence of detectable metastases can be a relatively slow process spreading over years or even decades [76, 133]. This extended period of latency suggests the establishment of tumor dormancy, which can occur in two different manners. Single cell dormancy – in which cells exit the cell cycle and are arrested in G_0 – should be distinguished from population-level dormancy in which cell proliferation is balanced with apoptosis. The latter occurs in response to immunosurveillance or when micrometastases fail to initiate angiogenesis [74, 125, 140]. Clinical analysis of bone marrow aspirates from asymptomatic breast and gastrointestinal cancer survivors revealed that the majority of DTCs are single cells or rarer small cell clusters that are quiescent, supporting the establishment of single cell dormancy in patients [112]. On the other hand, population level dormancy is supported by continuous shedding of CTCs in patients without overt metastases [142, 150].

2.3.2.1 Disseminated Tumor Cells Reside in Perivascular Niches

Once in a distant tissue, disseminated breast tumor cells settle in close vicinity of blood vessels, within the so-called perivascular niche (PVN) [63, 120]. The new microenvironment they encounter dictates their ability to balance between dormant and proliferative states [2]. For example, the extracellular signal-regulated kinase (Erk)/p38 ratio downstream of urokinase-type plasminogen activator receptor (uPAR)/$\alpha5\beta1$ integrin is a strong predictor of dormancy *in vivo*. Indeed, DTCs with high Erk activity tend to be proliferative, while high p38 activation forces cells into dormancy [3].

The PVN is a tissue-specific entity composed of endothelial cells, pericytes, fibroblasts, local immune cells, epithelial cells, stem and progenitor cells and extracellular matrix/basement membrane components [63]. The resident cell types vary between tissue microenvironments, e.g. in the brain, astrocytes and microglia are components of the PVN, whereas in liver, the PVN consists of stellate cells, hepatocytes and Kuppfer cells (resident liver macrophages) [19]. As described in the following examples, these different cell types influence DTC survival and dormancy. While it is not clear whether CSCs and dormant tumor cells are as one or whether they represent two distinct populations sharing the same niche, it is evident that they share similar gene expression signatures and might therefore respond similarly to microenvironmental cues [87]. Thus, additional examples extrapolated from normal and CSC quiescence are also presented.

TGF-β and bone morphogenic proteins (BMPs) represent central regulators of dormancy. In the bone marrow, stromal cells secrete BMP-7, which sustains prostate cancer cell dormancy through induction of the cell cycle inhibitors p21 and p27 [85]. Chronic stress has been associated with cancer relapse and norepinephrine secreted by the adrenergic neurons contributes to the reactivation of dormant prostate cancer cells by repressing growth arrest-specific gene (Gas)-6, a regulator of the dormancy-inducing pathway Axl-TGF-β1/2-TGFβR2/3, in osteoblasts [39, 161]. Single cell analysis of breast cancer patient samples also supports the implication of the TGFβ2-TGFβR3 axis in breast tumor cell dormancy [87], and experiments performed in a head and neck squamous cell carcinoma model identified the TGF-β2-p38 axis as a regulator of p27 expression [16]. Whereas most of the cell types present in within the PVN likely play some role in influencing DTC phenotype, endothelial cells are of particular interest in this process, particularly in regards to anti-angiogenic therapy.

2.3.2.2 Vascular Homeostasis Regulates Disseminated Tumor Cell Behavior

Long considered to be a passive actor in homeostasis by providing nutrients and oxygen to organs, the endothelium is now recognized as an essential player in tissue regeneration and homeostasis maintenance [21, 42, 43, 123]. As described below, endothelial cells and stem cells evolve hand in hand: resting endothelium secretes proteins responsible for inducing stem cell quiescence, whereas activated endothelium that is sprouting in response to tissue damage and inflammation secretes factors implicated in stem cell and progenitor cell expansion. It is therefore not surprising to find stem cells in close vicinity of the endothelium and one can easily anticipate endothelial cells to be involved in DTC survival, dormancy and reactivation.

Calabrese and collaborators demonstrated that nestin-positive brain CSCs are – like normal neural stem cells – found in close association with tumor blood vessels, which maintain CSC self-renewing capabilities [22]. In another study, intravital imaging in brain revealed that lung cancer and melanoma DTCs migrating away from the endothelium following extravasation regress, suggesting that endothelial cells support DTC survival [84]. A very small DTC subset remained dormant for 51 days and all of them were located in the PVN. Similarly, in chorioallantoic membrane assays and murine models of cancer, tumor cells can persist in a quiescent state despite the presence of a rich blood vessel network providing them with nutrients and oxygen [4]. In lung and bone marrow as well, the endothelium regulates DTC dormancy. Mature, homeostatic blood vessels secrete an array of factors, including thrombospondin-1, which maintain DTCs in a quiescent state [64]. Inversely, TGF-β1 and periostin secreted by sprouting neovasculature were shown to awaken tumor cells and induce their proliferation [64]. Supporting these data, bone marrow biopsies from breast cancer patients revealed that quiescent micrometastases localize to stromal cell-derived factor (SDF)-1-positive perisinusoidal niches [120]. Further analysis performed in mice suggested that breast cancer cells are mobilized to the bone marrow by E-selectin-expressing sinusoids, to which they anchor via SDF-1/CXC chemokine receptor (CXCR)-4 interactions post-extravasation. Detachment of these cells from the vasculature triggers their proliferation.

4T07 mammary cancer cells are found at distant sites like the lung, where they reside in close vicinity to blood vessels and are maintained in a quiescent state by endothelial-derived BMPs [59]. Induction of Coco, a BMP-4 antagonist, induces their exit from dormancy in a lung-specific fashion.

Similarly, T-cell acute lymphoblastic leukemia cells reside on the bone marrow endothelium and their expansion is directly linked to vascular CXC chemokine ligand (CXCL)-12 expression [118]. Targeting CXCR-4 – a CXCL-12 receptor – in the leukemic cells induces their apoptosis in the vessel lumen, leading to sustained remission. CXCL-12 has also been described as a lung-derived chemoattractant implicated in the establishment of lung metastasis and as an autocrine survival factor for glioblastoma CSCs [60, 116].

Supporting the necessity for DTCs to reside in close vicinity of the blood vessels in order to remain dormant, non-diffusible factors, such as the cell-cell communication pathway Notch or various extracellular matrix-related proteins have also been implicated in the regulation of CSC quiescence.

Notch ligands have opposing effects on quiescence. In the brain, Dll4- and Jagged1-expressing endothelial cells contribute to glioblastoma outgrowth and CSC renewal [162]. However, Notch signaling can also maintain quiescence, as shown in some brain tumor subtypes, such as astrocytomas or proneural glioblastomas [65]. Additionally, endothelial cells secrete a soluble version of Jagged1, which enriches for chemoresistant colon CSCs [91]. Loss of nitric oxide in the PVN is associated with repression of Notch signaling in glioma cells, leading to decreased stemness and prolonged survival of mice [29].

Extracellular matrix components and related proteins also regulate dormancy and outgrowth. For example, laminin α2 originating from the endothelium is known to induce glioblastoma stem cell growth [86]. Endothelial cell-derived insulin-like growth factor binding protein-(IGFBP)-7 inhibits cancer stem-like cell expansion by repressing insulin-like growth factor (IGF)-1-insulin-like growth factor-1 receptor (IGF1R) signaling, a mechanism reversed during chemotherapy, leading to enhanced progression [25].

Protected by the PVN, dormant DTCs are resistant to conventional chemotherapy. Identifying drugs able to maintain DTCs in a dormant state, or, even better, drugs able to definitively eradicate them without chemotherapy, would prevent long term recurrence in cancer patients [63]. Considering the importance of the PVN, and more precisely the endothelium, in regulation of DTC behavior, one may wonder whether disruption of this niche using anti-angiogenic therapy is a prudent approach.

2.3.3 Anti-angiogenics, Real Dormancy Inducers?

Paradoxically, whereas anti-angiogenic therapies were designed originally to maintain dormancy, their effect on metastatic niches in general and on DTCs in particular have been poorly studied. As just discussed, stable microvascular endothelium is a critical component of the dormant niche of disseminated breast tumor cells. However, anti-angiogenics also damage quiescent endothelium. So, how could anti-angiogenics impact the dormant tumor niche?

2.3.3.1 Direct Effect of Anti-angiogenic Therapy on Disseminated Tumor Cells

"Angiogenic" factors are not only essential to endothelial cell survival. Tumor cells and more particularly CSCs also rely on them in order to survive and proliferate. The most representative example is probably the implication of VEGF-A in CSC maintenance [67]. Using a chemical model of skin papillomas, Beck and colleagues demonstrated that VEGF-A plays a dual role in tumor initiation. By binding to endothelial VEGFR-2, VEGF-A contributes to the formation of the vascular niche in which CSCs reside. Interestingly, they also identified a VEGF-neuropilin1 loop, which directly supports the renewal potential of CSCs [12]. Similarly, Goel and collaborators observed that activation of neuropilin2-α6β1 integrin in response to VEGF-A decreases tumor-free survival in mice [68]. However, the role of these signaling pathways in CSC proliferation at metastatic sites was not assessed in the context of these studies. Nevertheless, the fact that VEGF regulates CSC maintenance through neuropilins rather than through VEGFRs suggests once again that targeting VEGF-A using bevacizumab might be more efficient than the use of tyrosine kinase inhibitors. This hypothesis is also supported by the increased levels of VEGF-A in the livers of mice treated with sunitinib, which could foster CSC renewal [47]. Combining extracellular VEGF/VEGFR inhibitors with FAK inhibitors or neuropilin-targeting peptides might therefore prove more efficient in eliminating DTCs – should the combination not reveal too toxic.

2.3.3.2 Targeting the Niche with Anti-angiogenic Therapy

As discussed in Sect. 2.3.2.2, endothelial cell-secreted factors promote breast tumor cell dormancy as long as endothelial homeostasis is maintained. This raises the question: Could disruption of vascular homeostasis induced by angiogenesis inhibitors reactivate dormant DTCs? Normal endothelial homeostasis is regulated by autocrine VEGF-VEGFR-2 signaling [88]. Thus, we suspect that tyrosine kinase inhibitors, such as sunitinib, will not only interfere with sprouting angiogenesis but also damage "normal" quiescent endothelium at distant sites, a hypothesis supported by reduced capillary density in the trachea or the liver of sunitinib-treated mice [31, 160].

Indeed, vascular cell adhesion molecule (VCAM)-1, an endothelial damage marker expressed in atherosclerosis and inflammatory diseases, is elevated in the serum of sunitinib-treated patients [98, 149]. In a mouse mammary tumor dormancy model, VCAM-1 expression induced the recruitment of osteoclast progenitors, which activated indolent micrometastases located in the bone marrow and resulted in early mammary cancer recurrence, suggesting that damaged endothelium might as well recruit osteoclast progenitors and induce DTC re-emergence [92].

Additionally, sunitinib treatment increases serum levels of inflammatory cytokines such as CXCL-12, stem cell factor (SCF) and granulocyte-colony stimulating factor (G-CSF) [47]. As inflammation in general and CXCL-12 in particular have recently been shown to promote DTC reactivation, it would not be surprising to observe DTC outgrowth following anti-angiogenic treatment [38, 118].

Finally, one could imagine that endothelial cell apoptosis following anti-angiogenic therapy would lead to a decreased expression of endothelial cell-derived dormancy regulators, such as thrombospondin-1 [64] (Fig. 2.1e, f). One should however not forget that endothelial cells do not only induce DTC dormancy: they also promote DTC survival; it is currently unknown which of these effects would be the most impacted by anti-angiogenics.

2.3.3.3 Discontinuation of Anti-angiogenic Therapy, the Calm After the Storm?

Sunitinib is given to breast cancer patients either continuously or on a 4-weeks-ON-2-weeks-OFF schedule, whereas bevacizumab is infused every second week [104, 145]. How do the metastatic niche and DTCs respond to therapeutic holiday? The following data suggest that it might actually have deleterious effects.

Using MDA-MB-231 and 4T1 cells, Sounni and colleagues showed that sunitinib-treated tumors become glycolytic, but they shift towards lipid synthesis after sunitinib withdrawal [141]. Both primary and metastatic tumor outgrowth are fostered following this metabolic reprogramming. Pharmaceutical and shRNA-mediated inhibition of fatty acid synthase interfered with primary tumor regrowth and metastatic formation after sunitinib withdrawal in breast and colon cancer models.

In a recent study, discontinuation of angiogenesis inhibitors was also associated with a 15–20-fold increase in the liver sinusoid pore size [160]. Three to six weeks were necessary before the pore size returned to normal and this period was characterized by a loss of VE-cadherin expression. As a result, colorectal carcinoma cells showed an increased ability to intra- and extravasate. Moreover, this work showed that liver revascularization is initiated 6 days post-treatment withdrawal and reaches its original level by day 12. Thus, between treatment cycles, distant organs are most likely to alternate between vascular damage and regeneration, both bringing their share of inflammatory and growth factors and putting the patient at risk of recurrence. Indeed, thymic, pulmonary and hepatic microvascular endothelial cells are central to tissue regeneration via the secretion of angiocrine factors such as BMP-4, matrix metalloproteinase (MMP)-14, hepatocyte growth factor (HGF) or Wnt2 [42, 43, 156]. Whereas BMP4 may sustain dormancy [59], the mitogenic effects of HGF and Wnt2 and the growth factors released by MMP-14 could induce their proliferation [93, 132, 144]. Finally, in the liver, ischemia/reperfusion – which is most likely to happen during these cycles of vascular damage and regeneration – has been associated with outgrowth of colon micrometastases [148].

2.4 Perspectives

This chapter sheds light on the dark side of anti-angiogenic therapies, exploring the hypothesis that they may actually promote metastatic spread. As described, targeting the vasculature is a risky bet, which can lead to increased tumor invasion, intra- and extra-vasation, and outgrowth at metastatic sites.

Thus, is anti-angiogenic therapy a valid option in preventing breast cancer recurrence long-term? It is not trivial to determine the potential of anti-angiogenic therapy in DTC eradication and whether targeting the dormant tumor niche using such approach would help prevent long-term recurrence as (i) most of the clinical data accumulated so far were in a context of already metastatic disease and (ii) we do not benefit from sufficient hindsight regarding the rare clinical trials performed in early-stage cancer. In fact, use of anti-angiogenics in neo-adjuvant settings has been associated with increased pathological complete response in stage II-III breast cancer patients, but their effect on metastasis-free survival and overall survival are still under evaluation [139].

What directions could we take to make anti-angiogenics more efficient at targeting not only the primary tumor, but also metastatic seeds.

1. Normalize vessels rather than destroy them. We have seen that VEGF inhibition, more especially when intracellular, does not only damage tumor infiltrating blood vessels but also damages mature blood vessels located at distant sites, leading to increased tumor cell dissemination. Favoring vessel normalization would pre-empt blood vessel recruitment with limited damage on pre-existing vessels [27]. Additionally, vessel normalization at the primary site would prevent hypoxia-induced invasiveness and CSC enrichment, and would in addition lead to more efficient chemotherapy delivery to the tumor [24, 113, 121].
2. Combine therapies. Combining angiogenesis inhibition with chemotherapy has been associated with a reduction of metastases when compared to angiogenesis inhibition alone [79, 111]. Alternately, combining VEGF and angiopoietin-2 inhibition could also prove useful, as it could simultaneously inhibit angiogenesis, vessel co-option and CSC renewal [83, 143]. Finally, considering that dormancy is regulated by vascular homeostasis and immune control https://www.ncbi.nlm.nih.gov/pubmed/27613129, combining vessel normalization and immunostimulation also offers great promise in preventing DTC outgrowth [6, 147].
3. Refine therapeutic schedule. Two options could be considered. First of all, anti-angiogenic therapy should probably be avoided in neo-adjuvant settings to prevent increased primary tumor invasiveness. And second, anti-angiogenics should be administered regularly enough (perhaps at lower doses) to stymie rebound angiogenesis during drug holidays [160].
4. Use non-VEGF-targeting anti-angiogenic therapy. As discussed here, not all metastases rely on VEGF-mediated angiogenesis. Would non-VEGF-targeted therapy prove more effective? This hypothesis has been tested by Seaman and collaborators [135]. The authors observed that CD276 is highly upregulated in tumor cells of most origins and is additionally expressed by tumor infiltrating endothelial cells. Based on this, they developed an

antibody-drug conjugate against CD276. When linked to pyrrolobenzodiazepine – a DNA crosslinking compound – this antibody-drug conjugate can kill both the tumor cells and associated endothelium. However, the activity of this conjugate against dormant DTCs and the ostensibly "normal" endothelium in the dormant niche is unknown. If, for instance, CD276 is expressed by freshly re-emerging DTCs, targeting this molecule could prove beneficial in preventing metastatic outgrowth with reduced side effects, since CD276 is not expressed by normal cells and non-tumor-associated endothelium.

Our main challenge remains the lack of knowledge regarding the effects of anti-angiogenic therapy at distant sites – on single DTCs, on micrometastases, and even on macrometastases and their activated surroundings [107]. A better understanding of the molecular mechanisms governing tumor dormancy, its regulation within distant tissue microenvironments, and its potential dependence on vascular homeostasis should inspire more efficient treatments able to prolong long-term survival of breast cancer patients.

Acknowledgments The authors would like to thank Dr. Candice Grzelak for providing feedback on the manuscript. L.P. is funded by the Swiss National Science Foundation (Postdoc.Mobility Fellowships 165389 and 177917). The Ghajar Laboratory is funded by the U.S. Department of Defense Breast Cancer Research Program (BCRP; W841XWH-15-1-0201), a Physical Sciences Oncology Project Grant from the NIH/NCI (U54CA193461-01), the Keck Foundation, and the Breast Cancer Research Foundation.

References

1. Aguilera KY, Rivera LB, Hur H, Carbon JG, Toombs JE, Goldstein CD, Dellinger MT, Castrillon DH, Brekken RA (2014) Collagen signaling enhances tumor progression after anti-VEGF therapy in a murine model of pancreatic ductal adenocarcinoma. Cancer Res 74(4):1032–1044. https://doi.org/10.1158/0008-5472.CAN-13-2800
2. Aguirre-Ghiso JA (2007) Models, mechanisms and clinical evidence for cancer dormancy. Nat Rev Cancer 7(11):834–846. https://doi.org/10.1038/nrc2256
3. Aguirre-Ghiso JA, Estrada Y, Liu D, Ossowski L (2003) ERK(MAPK) activity as a determinant of tumor growth and dormancy; regulation by p38(SAPK). Cancer Res 63(7):1684–1695
4. Aguirre-Ghiso JA, Ossowski L, Rosenbaum SK (2004) Green fluorescent protein tagging of extracellular signal-regulated kinase and p38 pathways reveals novel dynamics of pathway activation during primary and metastatic growth. Cancer Res 64(20):7336–7345. https://doi.org/10.1158/0008-5472.CAN-04-0113
5. Algire GH, Chalkley HW (1945) Vascular reactions of normal and malignant tissue in vivo. I Vascular reactions of mice to wounds and or normal and neoplastic transplants. J Natl Cancer Inst 6:73–85
6. Allen E, Jabouille A, Rivera LB, Lodewijckx I, Missiaen R, Steri V, Feyen K, Tawney J, Hanahan D, Michael IP, Bergers G (2017) Combined antiangiogenic and anti-PD-L1 therapy stimulates tumor immunity through HEV formation. Sci Transl Med 9(385). https://doi.org/10.1126/scitranslmed.aak9679
7. Allen E, Mieville P, Warren CM, Saghafinia S, Li L, Peng MW, Hanahan D (2016) Metabolic symbiosis enables adaptive resistance to anti-angiogenic therapy that is dependent on mTOR signaling. Cell Rep 15(6):1144–1160. https://doi.org/10.1016/j.celrep.2016.04.029
8. Ang YL, Soo RA, Ho GF, Sundar R, Tan SH, Lim JSJ, Yong W-P, Chong WQ, Ho J, Kumarakulasinghe NB, Aung MI, Koh E, Ramlee M, Phyu PS, Kyin L, Goh BC, Lee S-C (2016) A phase II randomized study of docetaxel +/- low-dose, short course sunitinib in advanced solid tumours. J Clin Oncol 34(15_suppl):e14124–e14124. https://doi.org/10.1200/JCO.2016.34.15_suppl.e14124
9. Banys M, Solomayer EF, Gebauer G, Janni W, Krawczyk N, Lueck HJ, Becker S, Huober J, Kraemer B, Wackwitz B, Hirnle P, Wallwiener D, Fehm T (2013) Influence of zoledronic acid on disseminated tumor cells in bone marrow and survival: results of a prospective clinical trial. BMC Cancer 13:480. https://doi.org/10.1186/1471-2407-13-480
10. Bear HD, Tang G, Rastogi P, Geyer CE Jr, Liu Q, Robidoux A, Baez-Diaz L, Brufsky AM, Mehta RS, Fehrenbacher L, Young JA, Senecal FM, Gaur R, Margolese RG, Adams PT, Gross HM, Costantino JP, Paik S, Swain SM, Mamounas EP, Wolmark N (2015) Neoadjuvant plus adjuvant bevacizumab in early breast cancer (NSABP B-40 [NRG Oncology]): secondary outcomes of a phase 3, randomised controlled trial. The Lancet Oncol 16(9):1037–1048. https://doi.org/10.1016/S1470-2045(15)00041-8
11. Bear HD, Tang G, Rastogi P, Geyer CE Jr, Robidoux A, Atkins JN, Baez-Diaz L, Brufsky AM, Mehta RS, Fehrenbacher L, Young JA, Senecal FM, Gaur R, Margolese RG, Adams PT, Gross HM, Costantino JP, Swain SM, Mamounas EP, Wolmark N (2012)

Bevacizumab added to neoadjuvant chemotherapy for breast cancer. N Engl J Med 366(4):310–320. https://doi.org/10.1056/NEJMoa1111097

12. Beck B, Driessens G, Goossens S, Youssef KK, Kuchnio A, Caauwe A, Sotiropoulou PA, Loges S, Lapouge G, Candi A, Mascre G, Drogat B, Dekoninck S, Haigh JJ, Carmeliet P, Blanpain C (2011) A vascular niche and a VEGF-Nrp1 loop regulate the initiation and stemness of skin tumours. Nature 478(7369):399–403. https://doi.org/10.1038/nature10525
13. Bell R, Brown J, Parmar M, Toi M, Suter T, Steger GG, Pivot X, Mackey J, Jackisch C, Dent R, Hall P, Xu N, Morales L, Provencher L, Hegg R, Vanlemmens L, Kirsch A, Schneeweiss A, Masuda N, Overkamp F, Cameron D (2017) Final efficacy and updated safety results of the randomized phase III BEATRICE trial evaluating adjuvant bevacizumab-containing therapy in triple-negative early breast cancer. Ann Oncol 28(4):754–760
14. Bergers G, Hanahan D (2008) Modes of resistance to anti-angiogenic therapy. Nat Rev Cancer 8(8):592–603. https://doi.org/10.1038/nrc2442
15. Bergh J, Bondarenko IM, Lichinitser MR, Liljegren A, Greil R, Voytko NL, Makhson AN, Cortes J, Lortholary A, Bischoff J, Chan A, Delaloge S, Huang X, Kern KA, Giorgetti C (2012) First-line treatment of advanced breast cancer with sunitinib in combination with docetaxel versus docetaxel alone: results of a prospective, randomized phase III study. J Clin Oncol 30(9):921–929. https://doi.org/10.1200/JCO.2011.35.7376
16. Bragado P, Estrada Y, Parikh F, Krause S, Capobianco C, Farina HG, Schewe DM, Aguirre-Ghiso JA (2013) TGF-beta2 dictates disseminated tumour cell fate in target organs through TGF-beta-RIII and p38alpha/beta signalling. Nat Cell Biol 15(11):1351–1361. https://doi.org/10.1038/ncb2861
17. Braun S, Vogl FD, Naume B, Janni W, Osborne MP, Coombes RC, Schlimok G, Diel IJ, Gerber B, Gebauer G, Pierga JY, Marth C, Oruzio D, Wiedswang G, Solomayer EF, Kundt G, Strobl B, Fehm T, Wong GY, Bliss J, Vincent-Salomon A, Pantel K (2005) A pooled analysis of bone marrow micrometastasis in breast cancer. N Engl J Med 353(8):793–802. https://doi.org/10.1056/NEJMoa050434
18. Bridgeman VL, Vermeulen PB, Foo S, Bilecz A, Daley F, Kostaras E, Nathan MR, Wan E, Frentzas S, Schweiger T, Hegedus B, Hoetzenecker K, Renyi-Vamos F, Kuczynski EA, Vasudev NS, Larkin J, Gore M, Dvorak HF, Paku S, Kerbel RS, Dome B, Reynolds AR (2017) Vessel co-option is common in human lung metastases and mediates resistance to anti-angiogenic therapy in preclinical lung metastasis models. J Pathol 241(3):362–374. https://doi.org/10.1002/path.4845
19. Brodt P (2016) Role of the microenvironment in liver metastasis: from pre- to prometastatic niches. Clin Cancer Res 22(24):5971–5982. https://doi.org/10.1158/1078-0432.CCR-16-0460
20. Brufsky AM, Hurvitz S, Perez E, Swamy R, Valero V, O'Neill V, Rugo HS (2011) RIBBON-2: a randomized, double-blind, placebo-controlled, phase III trial evaluating the efficacy and safety of bevacizumab in combination with chemotherapy for second-line treatment of human epidermal growth factor receptor 2-negative metastatic breast cancer. J Clin Oncol 29(32):4286–4293. https://doi.org/10.1200/JCO.2010.34.1255
21. Butler JM, Kobayashi H, Rafii S (2010) Instructive role of the vascular niche in promoting tumour growth and tissue repair by angiocrine factors. Nat Rev Cancer 10(2):138–146. https://doi.org/10.1038/nrc2791
22. Calabrese C, Poppleton H, Kocak M, Hogg TL, Fuller C, Hamner B, Oh EY, Gaber MW, Finklestein D, Allen M, Frank A, Bayazitov IT, Zakharenko SS, Gajjar A, Davidoff A, Gilbertson RJ (2007) A perivascular niche for brain tumor stem cells. Cancer Cell 11(1):69–82. https://doi.org/10.1016/j.ccr.2006.11.020
23. Cameron D, Brown J, Dent R, Jackisch C, Mackey J, Pivot X, Steger GG, Suter TM, Toi M, Parmar M, Laeufle R, Im Y-H, Romieu G, Harvey V, Lipatov O, Pienkowski T, Cottu P, Chan A, Im S-A, Hall PS, Bubuteishvili-Pacaud L, Henschel V, Deurloo RJ, Pallaud C, Bell R (2013) Adjuvant bevacizumab-containing therapy in triple-negative breast cancer (BEATRICE): primary results of a randomised, phase 3 trial. The Lancet Oncol 14(10):933–942
24. Cantelmo AR, Conradi LC, Brajic A, Goveia J, Kalucka J, Pircher A, Chaturvedi P, Hol J, Thienpont B, Teuwen LA, Schoors S, Boeckx B, Vriens J, Kuchnio A, Veys K, Cruys B, Finotto L, Treps L, Stav-Noraas TE, Bifari F, Stapor P, Decimo I, Kampen K, De Bock K, Haraldsen G, Schoonjans L, Rabelink T, Eelen G, Ghesquiere B, Rehman J, Lambrechts D, Malik AB, Dewerchin M, Carmeliet P (2016) Inhibition of the glycolytic activator PFKFB3 in endothelium induces tumor vessel normalization, impairs metastasis, and improves chemotherapy. Cancer Cell 30(6):968–985. https://doi.org/10.1016/j.ccell.2016.10.006
25. Cao Z, Scandura JM, Inghirami GG, Shido K, Ding BS, Rafii S (2017) Molecular checkpoint decisions made by subverted vascular niche transform indolent tumor cells into chemoresistant cancer stem cells. Cancer Cell 31(1):110–126. https://doi.org/10.1016/j.ccell.2016.11.010
26. Carmeliet P (2005) Angiogenesis in life, disease and medicine. Nature 438(7070):932–936. https://doi.org/10.1038/nature04478
27. Carmeliet P, Jain RK (2011) Principles and mechanisms of vessel normalization for cancer and other angiogenic diseases. Nat Rev Drug Discov 10(6):417–427. https://doi.org/10.1038/nrd3455
28. Casillas AL, Toth RK, Sainz AG, Singh N, Desai AA, Kraft AS, Warfel NA (2017) Hypoxia-inducible

PIM kinase expression promotes resistance to anti-angiogenic agents. Clin Cancer Res. https://doi.org/10.1158/1078-0432.CCR-17-1318
29. Charles N, Ozawa T, Squatrito M, Bleau AM, Brennan CW, Hambardzumyan D, Holland EC (2010) Perivascular nitric oxide activates notch signaling and promotes stem-like character in PDGF-induced glioma cells. Cell Stem Cell 6(2):141–152. https://doi.org/10.1016/j.stem.2010.01.001
30. Christensen JG (2007) A preclinical review of sunitinib, a multitargeted receptor tyrosine kinase inhibitor with anti-angiogenic and antitumour activities. Ann Oncol 18(Suppl 10):x3–10. https://doi.org/10.1093/annonc/mdm408
31. Chung AS, Kowanetz M, Wu X, Zhuang G, Ngu H, Finkle D, Komuves L, Peale F, Ferrara N (2012) Differential drug class-specific metastatic effects following treatment with a panel of angiogenesis inhibitors. J Pathol 227(4):404–416. https://doi.org/10.1002/path.4052
32. Cohen MH, Shen YL, Keegan P, Pazdur R (2009) FDA drug approval summary: bevacizumab (Avastin) as treatment of recurrent glioblastoma multiforme. Oncologist 14(11):1131–1138. https://doi.org/10.1634/theoncologist.2009-0121
33. Conley SJ, Baker TL, Burnett JP, Theisen RL, Lazarus D, Peters CG, Clouthier SG, Eliasof S, Wicha MS (2015) CRLX101, an investigational camptothecin-containing nanoparticle-drug conjugate, targets cancer stem cells and impedes resistance to antiangiogenic therapy in mouse models of breast cancer. Breast Cancer Res Treat 150(3):559–567. https://doi.org/10.1007/s10549-015-3349-8
34. Conley SJ, Gheordunescu E, Kakarala P, Newman B, Korkaya H, Heath AN, Clouthier SG, Wicha MS (2012) Antiangiogenic agents increase breast cancer stem cells via the generation of tumor hypoxia. Proc Natl Acad Sci U S A 109(8):2784–2789. https://doi.org/10.1073/pnas.1018866109
35. Cooke VG, LeBleu VS, Keskin D, Khan Z, O'Connell JT, Teng Y, Duncan MB, Xie L, Maeda G, Vong S, Sugimoto H, Rocha RM, Damascena A, Brentani RR, Kalluri R (2012) Pericyte depletion results in hypoxia-associated epithelial-to-mesenchymal transition and metastasis mediated by met signaling pathway. Cancer Cell 21(1):66–81. https://doi.org/10.1016/j.ccr.2011.11.024
36. Crown JP, Dieras V, Staroslawska E, Yardley DA, Bachelot T, Davidson N, Wildiers H, Fasching PA, Capitain O, Ramos M, Greil R, Cognetti F, Fountzilas G, Blasinska-Morawiec M, Liedtke C, Kreienberg R, Miller WH Jr, Tassell V, Huang X, Paolini J, Kern KA, Romieu G (2013) Phase III trial of sunitinib in combination with capecitabine versus capecitabine monotherapy for the treatment of patients with pretreated metastatic breast cancer. J Clin Oncol 31(23):2870–2878. https://doi.org/10.1200/JCO.2012.43.3391
37. Curigliano G, Pivot X, Cortes J, Elias A, Cesari R, Khosravan R, Collier M, Huang X, Cataruozolo PE, Kern KA, Goldhirsch A (2013) Randomized phase II study of sunitinib versus standard of care for patients with previously treated advanced triple-negative breast cancer. Breast (Edinburgh, Scotland) 22(5):650–656
38. De Cock JM, Shibue T, Dongre A, Keckesova Z, Reinhardt F, Weinberg RA (2016) Inflammation triggers Zeb1-dependent escape from tumor latency. Cancer Res 76(23):6778–6784. https://doi.org/10.1158/0008-5472.CAN-16-0608
39. Decker AM, Jung Y, Cackowski FC, Yumoto K, Wang J, Taichman RS (2017) Sympathetic signaling reactivates quiescent disseminated prostate cancer cells in the bone marrow. Mol Cancer Res: MCR 15(12):1644–1655. https://doi.org/10.1158/1541-7786.MCR-17-0132
40. Dickler MN, Barry WT, Cirrincione CT, Ellis MJ, Moynahan ME, Innocenti F, Hurria A, Rugo HS, Lake DE, Hahn O, Schneider BP, Tripathy D, Carey LA, Winer EP, Hudis CA (2016) Phase III trial evaluating letrozole as first-line endocrine therapy with or without bevacizumab for the treatment of postmenopausal women with hormone receptor-positive advanced-stage breast cancer: CALGB 40503 (Alliance). J Clin Oncol 34(22):2602–2609. https://doi.org/10.1200/JCO.2015.66.1595
41. Dieras V, Wildiers H, Jassem J, Dirix LY, Guastalla JP, Bono P, Hurvitz SA, Goncalves A, Romieu G, Limentani SA, Jerusalem G, Lakshmaiah KC, Roche H, Sanchez-Rovira P, Pienkowski T, Segui Palmer MA, Li A, Sun YN, Pickett CA, Slamon DJ (2015) Trebananib (AMG 386) plus weekly paclitaxel with or without bevacizumab as first-line therapy for HER2-negative locally recurrent or metastatic breast cancer: A phase 2 randomized study. Breast 24(3):182–190. https://doi.org/10.1016/j.breast.2014.11.003
42. Ding BS, Nolan DJ, Butler JM, James D, Babazadeh AO, Rosenwaks Z, Mittal V, Kobayashi H, Shido K, Lyden D, Sato TN, Rabbany SY, Rafii S (2010) Inductive angiocrine signals from sinusoidal endothelium are required for liver regeneration. Nature 468(7321):310–315. https://doi.org/10.1038/nature09493
43. Ding BS, Nolan DJ, Guo P, Babazadeh AO, Cao Z, Rosenwaks Z, Crystal RG, Simons M, Sato TN, Worgall S, Shido K, Rabbany SY, Rafii S (2011) Endothelial-derived angiocrine signals induce and sustain regenerative lung alveolarization. Cell 147(3):539–553. https://doi.org/10.1016/j.cell.2011.10.003
44. Du R, Lu KV, Petritsch C, Liu P, Ganss R, Passegue E, Song H, Vandenberg S, Johnson RS, Werb Z, Bergers G (2008) HIF1alpha induces the recruitment of bone marrow-derived vascular modulatory cells to regulate tumor angiogenesis and invasion. Cancer Cell 13(3):206–220. https://doi.org/10.1016/j.ccr.2008.01.034
45. Earl HM, Hiller L, Dunn JA, Blenkinsop C, Grybowicz L, Vallier AL, Gounaris I, Abraham JE,

Hughes-Davies L, McAdam K, Chan S, Ahmad R, Hickish T, Rea D, Caldas C, Bartlett JMS, Cameron DA, Provenzano E, Thomas J, Hayward RL, Group ARI (2017) Disease-free and overall survival at 3.5 years for neoadjuvant bevacizumab added to docetaxel followed by fluorouracil, epirubicin and cyclophosphamide, for women with HER2 negative early breast cancer: ARTemis Trial. Ann Oncol 28(8):1817–1824. https://doi.org/10.1093/annonc/mdx173
46. Ebos JM, Kerbel RS (2011) Antiangiogenic therapy: impact on invasion, disease progression, and metastasis. Nat Rev Clin Oncol 8(4):210–221. https://doi.org/10.1038/nrclinonc.2011.21
47. Ebos JM, Lee CR, Christensen JG, Mutsaers AJ, Kerbel RS (2007) Multiple circulating proangiogenic factors induced by sunitinib malate are tumor-independent and correlate with antitumor efficacy. Proc Natl Acad Sci U S A 104(43):17069–17074. https://doi.org/10.1073/pnas.0708148104
48. Ebos JM, Lee CR, Cruz-Munoz W, Bjarnason GA, Christensen JG, Kerbel RS (2009) Accelerated metastasis after short-term treatment with a potent inhibitor of tumor angiogenesis. Cancer Cell 15(3):232–239. https://doi.org/10.1016/j.ccr.2009.01.021
49. Ebos JM, Mastri M, Lee CR, Tracz A, Hudson JM, Attwood K, Cruz-Munoz WR, Jedeszko C, Burns P, Kerbel RS (2014) Neoadjuvant antiangiogenic therapy reveals contrasts in primary and metastatic tumor efficacy. EMBO Mol Med 6(12):1561–1576. https://doi.org/10.15252/emmm.201403989
50. Ehrmann RL, Knoth M (1968) Choriocarcinoma. Transfilter stimulation of vasoproliferation in the hamster cheek pouch. Studied by light and electron microscopy. J Natl Cancer Inst 41(6):1329–1341
51. Elice F, Rodeghiero F (2010) Bleeding complications of antiangiogenic therapy: pathogenetic mechanisms and clinical impact. Thromb Res 125(Suppl 2):S55–S57. https://doi.org/10.1016/S0049-3848(10)70014-1
52. Escudier B, Eisen T, Stadler WM, Szczylik C, Oudard S, Staehler M, Negrier S, Chevreau C, Desai AA, Rolland F, Demkow T, Hutson TE, Gore M, Anderson S, Hofilena G, Shan M, Pena C, Lathia C, Bukowski RM (2009) Sorafenib for treatment of renal cell carcinoma: Final efficacy and safety results of the phase III treatment approaches in renal cancer global evaluation trial. J Clin Oncol 27(20):3312–3318. https://doi.org/10.1200/JCO.2008.19.5511
53. Escudier B, Pluzanska A, Koralewski P, Ravaud A, Bracarda S, Szczylik C, Chevreau C, Filipek M, Melichar B, Bajetta E, Gorbunova V, Bay JO, Bodrogi I, Jagiello-Gruszfeld A, Moore N, AT Investigators (2007) Bevacizumab plus interferon alfa-2a for treatment of metastatic renal cell carcinoma: a randomised, double-blind phase III trial. Lancet 370(9605):2103–2111. https://doi.org/10.1016/S0140-6736(07)61904-7
54. Fantozzi A, Gruber DC, Pisarsky L, Heck C, Kunita A, Yilmaz M, Meyer-Schaller N, Cornille K, Hopfer U, Bentires-Alj M, Christofori G (2014) VEGF-mediated angiogenesis links EMT-induced cancer stemness to tumor initiation. Cancer Res 74(5):1566–1575. https://doi.org/10.1158/0008-5472.CAN-13-1641
55. Ferrara N, Hillan KJ, Gerber HP, Novotny W (2004) Discovery and development of bevacizumab, an anti-VEGF antibody for treating cancer. Nat Rev Drug Discov 3(5):391–400. https://doi.org/10.1038/nrd1381
56. Folkman J (1971) Tumor angiogenesis: therapeutic implications. N Engl J Med 285(21):1182–1186. https://doi.org/10.1056/NEJM197111182852108
57. Franco M, Roswall P, Cortez E, Hanahan D, Pietras K (2011) Pericytes promote endothelial cell survival through induction of autocrine VEGF-A signaling and Bcl-w expression. Blood 118(10):2906–2917. https://doi.org/10.1182/blood-2011-01-331694
58. Frentzas S, Simoneau E, Bridgeman VL, Vermeulen PB, Foo S, Kostaras E, Nathan M, Wotherspoon A, Gao ZH, Shi Y, Van den Eynden G, Daley F, Peckitt C, Tan X, Salman A, Lazaris A, Gazinska P, Berg TJ, Eltahir Z, Ritsma L, Van Rheenen J, Khashper A, Brown G, Nystrom H, Sund M, Van Laere S, Loyer E, Dirix L, Cunningham D, Metrakos P, Reynolds AR (2016) Vessel co-option mediates resistance to anti-angiogenic therapy in liver metastases. Nat Med 22(11):1294–1302. https://doi.org/10.1038/nm.4197
59. Gao H, Chakraborty G, Lee-Lim AP, Mo Q, Decker M, Vonica A, Shen R, Brogi E, Brivanlou AH, Giancotti FG (2012) The BMP inhibitor Coco reactivates breast cancer cells at lung metastatic sites. Cell 150(4):764–779. https://doi.org/10.1016/j.cell.2012.06.035
60. Gatti M, Pattarozzi A, Bajetto A, Wurth R, Daga A, Fiaschi P, Zona G, Florio T, Barbieri F (2013) Inhibition of CXCL12/CXCR4 autocrine/paracrine loop reduces viability of human glioblastoma stem-like cells affecting self-renewal activity. Toxicology 314(2-3):209–220. https://doi.org/10.1016/j.tox.2013.10.003
61. Gerber HP, Ferrara N (2005) Pharmacology and pharmacodynamics of bevacizumab as monotherapy or in combination with cytotoxic therapy in preclinical studies. Cancer Res 65(3):671–680
62. Gerhardt H, Semb H (2008) Pericytes: gatekeepers in tumour cell metastasis? J Mol Med (Berl) 86(2):135–144. https://doi.org/10.1007/s00109-007-0258-2
63. Ghajar CM (2015) Metastasis prevention by targeting the dormant niche. Nat Rev Cancer 15(4):238–247. https://doi.org/10.1038/nrc3910
64. Ghajar CM, Peinado H, Mori H, Matei IR, Evason KJ, Brazier H, Almeida D, Koller A, Hajjar KA, Stainier DY, Chen EI, Lyden D, Bissell MJ (2013) The perivascular niche regulates breast tumour dormancy. Nat Cell Biol 15(7):807–817. https://doi.org/10.1038/ncb2767

65. Giachino C, Boulay JL, Ivanek R, Alvarado A, Tostado C, Lugert S, Tchorz J, Coban M, Mariani L, Bettler B, Lathia J, Frank S, Pfister S, Kool M, Taylor V (2015) A tumor suppressor function for notch signaling in forebrain tumor subtypes. Cancer Cell 28(6):730–742. https://doi.org/10.1016/j.ccell.2015.10.008
66. Gianni L, Romieu GH, Lichinitser M, Serrano SV, Mansutti M, Pivot X, Mariani P, Andre F, Chan A, Lipatov O, Chan S, Wardley A, Greil R, Moore N, Prot S, Pallaud C, Semiglazov V (2013) AVEREL: a randomized phase III Trial evaluating bevacizumab in combination with docetaxel and trastuzumab as first-line therapy for HER2-positive locally recurrent/metastatic breast cancer. J Clin Oncol 31(14):1719–1725. https://doi.org/10.1200/JCO.2012.44.7912
67. Goel HL, Mercurio AM (2013) VEGF targets the tumour cell. Nat Rev Cancer 13(12):871–882. https://doi.org/10.1038/nrc3627
68. Goel HL, Pursell B, Chang C, Shaw LM, Mao J, Simin K, Kumar P, Vander Kooi CW, Shultz LD, Greiner DL, Norum JH, Toftgard R, Kuperwasser C, Mercurio AM (2013) GLI1 regulates a novel neuropilin-2/alpha6beta1 integrin based autocrine pathway that contributes to breast cancer initiation. EMBO Mol Med 5(4):488–508. https://doi.org/10.1002/emmm.201202078
69. Greenblatt M, Shubi P (1968) Tumor angiogenesis: transfilter diffusion studies in the hamster by the transparent chamber technique. J Natl Cancer Inst 41(1):111–124
70. Greene HS (1941) Heterologous transplantation of mammalian tumors: I. The Transfer of rabbit tumors to alien species. J Exp Med 73(4):461–474
71. Guerin E, Man S, Xu P, Kerbel RS (2013) A model of postsurgical advanced metastatic breast cancer more accurately replicates the clinical efficacy of antiangiogenic drugs. Cancer Res 73(9):2743–2748. https://doi.org/10.1158/0008-5472.CAN-12-4183
72. Hanahan D, Folkman J (1996) Patterns and emerging mechanisms of the angiogenic switch during tumorigenesis. Cell 86(3):353–364
73. Hanahan D, Weinberg RA (2000) The hallmarks of cancer. Cell 100(1):57–70
74. Holmgren L, O'Reilly MS, Folkman J (1995) Dormancy of micrometastases: balanced proliferation and apoptosis in the presence of angiogenesis suppression. Nat Med 1(2):149–153
75. Hurvitz SA, Perez EA, Yamamoto H, Valero V, O'Neill V, Rugo HS (2012) Final overall survival (OS) and safety analyses of RIBBON-2, a randomized phase III trial of bevacizumab (BEV) versus placebo (PL) combined with second-line chemotherapy (CT) for HER2-negative BEV-naive metastatic breast cancer (MBC). J Clin Oncol 30(27_suppl):100
76. Husemann Y, Geigl JB, Schubert F, Musiani P, Meyer M, Burghart E, Forni G, Eils R, Fehm T, Riethmuller G, Klein CA (2008) Systemic spread is an early step in breast cancer. Cancer Cell 13(1):58–68. https://doi.org/10.1016/j.ccr.2007.12.003
77. Ide AG, Baker NH, Warren SL (1939) Vascularization of the Brown-Pearce rabbit epithelioma transplant as seen in the transparent ear chamber. Am J Roentgenol 42:891–899
78. Inai T, Mancuso M, Hashizume H, Baffert F, Haskell A, Baluk P, Hu-Lowe DD, Shalinsky DR, Thurston G, Yancopoulos GD, McDonald DM (2004) Inhibition of vascular endothelial growth factor (VEGF) signaling in cancer causes loss of endothelial fenestrations, regression of tumor vessels, and appearance of basement membrane ghosts. Am J Pathol 165(1):35–52. https://doi.org/10.1016/S0002-9440(10)63273-7
79. Jedeszko C, Paez-Ribes M, Di Desidero T, Man S, Lee CR, Xu P, Bjarnason GA, Bocci G, Kerbel RS (2015) Postsurgical adjuvant or metastatic renal cell carcinoma therapy models reveal potent antitumor activity of metronomic oral topotecan with pazopanib. Sci Transl Med 7(282):282ra250. https://doi.org/10.1126/scitranslmed.3010722
80. Jimenez-Valerio G, Martinez-Lozano M, Bassani N, Vidal A, Ochoa-de-Olza M, Suarez C, Garcia-Del-Muro X, Carles J, Vinals F, Graupera M, Indraccolo S, Casanovas O (2016) Resistance to antiangiogenic therapies by metabolic symbiosis in renal cell carcinoma PDX models and patients. Cell Rep 15(6):1134–1143. https://doi.org/10.1016/j.celrep.2016.04.015
81. Johnson DH, Fehrenbacher L, Novotny WF, Herbst RS, Nemunaitis JJ, Jablons DM, Langer CJ, DeVore RF 3rd, Gaudreault J, Damico LA, Holmgren E, Kabbinavar F (2004) Randomized phase II trial comparing bevacizumab plus carboplatin and paclitaxel with carboplatin and paclitaxel alone in previously untreated locally advanced or metastatic non-small-cell lung cancer. J Clin Oncol 22(11):2184–2191. https://doi.org/10.1200/JCO.2004.11.022
82. Keunen O, Johansson M, Oudin A, Sanzey M, Rahim SA, Fack F, Thorsen F, Taxt T, Bartos M, Jirik R, Miletic H, Wang J, Stieber D, Stuhr L, Moen I, Rygh CB, Bjerkvig R, Niclou SP (2011) Anti-VEGF treatment reduces blood supply and increases tumor cell invasion in glioblastoma. Proc Natl Acad Sci U S A 108(9):3749–3754. https://doi.org/10.1073/pnas.1014480108
83. Kienast Y, Klein C, Scheuer W, Raemsch R, Lorenzon E, Bernicke D, Herting F, Yu S, The HH, Martarello L, Gassner C, Stubenrauch KG, Munro K, Augustin HG, Thomas M (2013) Ang-2-VEGF-A CrossMab, a novel bispecific human IgG1 antibody blocking VEGF-A and Ang-2 functions simultaneously, mediates potent antitumor, antiangiogenic, and antimetastatic efficacy. Clin Cancer Res 19(24):6730–6740. https://doi.org/10.1158/1078-0432.CCR-13-0081
84. Kienast Y, von Baumgarten L, Fuhrmann M, Klinkert WE, Goldbrunner R, Herms J, Winkler F (2010) Real-time imaging reveals the single steps of

brain metastasis formation. Nat Med 16(1):116–122. https://doi.org/10.1038/nm.2072

85. Kobayashi A, Okuda H, Xing F, Pandey PR, Watabe M, Hirota S, Pai SK, Liu W, Fukuda K, Chambers C, Wilber A, Watabe K (2011) Bone morphogenetic protein 7 in dormancy and metastasis of prostate cancer stem-like cells in bone. J Exp Med 208(13):2641–2655. https://doi.org/10.1084/jem.20110840
86. Lathia JD, Li M, Hall PE, Gallagher J, Hale JS, Wu Q, Venere M, Levy E, Rani MR, Huang P, Bae E, Selfridge J, Cheng L, Guvenc H, McLendon RE, Nakano I, Sloan AE, Phillips HS, Lai A, Gladson CL, Bredel M, Bao S, Hjelmeland AB, Rich JN (2012) Laminin alpha 2 enables glioblastoma stem cell growth. Ann Neurol 72(5):766–778. https://doi.org/10.1002/ana.23674
87. Lawson DA, Bhakta NR, Kessenbrock K, Prummel KD, Yu Y, Takai K, Zhou A, Eyob H, Balakrishnan S, Wang CY, Yaswen P, Goga A, Werb Z (2015) Single-cell analysis reveals a stem-cell program in human metastatic breast cancer cells. Nature 526(7571):131–135. https://doi.org/10.1038/nature15260
88. Lee S, Chen TT, Barber CL, Jordan MC, Murdock J, Desai S, Ferrara N, Nagy A, Roos KP, Iruela-Arispe ML (2007) Autocrine VEGF signaling is required for vascular homeostasis. Cell 130(4):691–703. https://doi.org/10.1016/j.cell.2007.06.054
89. Leung DW, Cachianes G, Kuang WJ, Goeddel DV, Ferrara N (1989) Vascular endothelial growth factor is a secreted angiogenic mitogen. Science 246(4935):1306–1309
90. Llovet JM, Ricci S, Mazzaferro V, Hilgard P, Gane E, Blanc JF, de Oliveira AC, Santoro A, Raoul JL, Forner A, Schwartz M, Porta C, Zeuzem S, Bolondi L, Greten TF, Galle PR, Seitz JF, Borbath I, Haussinger D, Giannaris T, Shan M, Moscovici M, Voliotis D, Bruix J, Group SIS (2008) Sorafenib in advanced hepatocellular carcinoma. N Engl J Med 359(4):378–390. https://doi.org/10.1056/NEJMoa0708857
91. Lu J, Ye X, Fan F, Xia L, Bhattacharya R, Bellister S, Tozzi F, Sceusi E, Zhou Y, Tachibana I, Maru DM, Hawke DH, Rak J, Mani SA, Zweidler-McKay P, Ellis LM (2013) Endothelial cells promote the colorectal cancer stem cell phenotype through a soluble form of Jagged-1. Cancer Cell 23(2):171–185. https://doi.org/10.1016/j.ccr.2012.12.021
92. Lu X, Mu E, Wei Y, Riethdorf S, Yang Q, Yuan M, Yan J, Hua Y, Tiede BJ, Lu X, Haffty BG, Pantel K, Massague J, Kang Y (2011) VCAM-1 promotes osteolytic expansion of indolent bone micrometastasis of breast cancer by engaging alpha4beta1-positive osteoclast progenitors. Cancer Cell 20(6):701–714. https://doi.org/10.1016/j.ccr.2011.11.002
93. Malladi S, Macalinao DG, Jin X, He L, Basnet H, Zou Y, de Stanchina E, Massague J (2016) Metastatic latency and immune evasion through autocrine inhibition of WNT. Cell 165(1):45–60. https://doi.org/10.1016/j.cell.2016.02.025
94. Mani SA, Guo W, Liao MJ, Eaton EN, Ayyanan A, Zhou AY, Brooks M, Reinhard F, Zhang CC, Shipitsin M, Campbell LL, Polyak K, Brisken C, Yang J, Weinberg RA (2008) The epithelial-mesenchymal transition generates cells with properties of stem cells. Cell 133(4):704–715. https://doi.org/10.1016/j.cell.2008.03.027
95. Marches R, Scheuermann R, Uhr J (2006) Cancer dormancy: from mice to man. Cell Cycle 5(16):1772–1778
96. Martin M, Loibl S, von Minckwitz G, Morales S, Martinez N, Guerrero A, Anton A, Aktas B, Schoenegg W, Munoz M, Garcia-Saenz JA, Gil M, Ramos M, Margeli M, Carrasco E, Liedtke C, Wachsmann G, Mehta K, De la Haba-Rodriguez JR (2015) Phase III trial evaluating the addition of bevacizumab to endocrine therapy as first-line treatment for advanced breast cancer: the letrozole/fulvestrant and avastin (LEA) study. J Clin Oncol 33(9):1045–1052. https://doi.org/10.1200/JCO.2014.57.2388
97. Martin M, Roche H, Pinter T, Crown J, Kennedy MJ, Provencher L, Priou F, Eiermann W, Adrover E, Lang I, Ramos M, Latreille J, Jagiello-Gruszfeld A, Pienkowski T, Alba E, Snyder R, Almel S, Rolski J, Munoz M, Moroose R, Hurvitz S, Banos A, Adewoye H, Hei YJ, Lindsay MA, Rupin M, Cabaribere D, Lemmerick Y, Mackey JR, T Investigators (2011) Motesanib, or open-label bevacizumab, in combination with paclitaxel, as first-line treatment for HER2-negative locally recurrent or metastatic breast cancer: a phase 2, randomised, double-blind, placebo-controlled study. Lancet Oncol 12(4):369–376. https://doi.org/10.1016/S1470-2045(11)70037-7
98. Marui N, Offermann MK, Swerlick R, Kunsch C, Rosen CA, Ahmad M, Alexander RW, Medford RM (1993) Vascular cell adhesion molecule-1 (VCAM-1) gene transcription and expression are regulated through an antioxidant-sensitive mechanism in human vascular endothelial cells. J Clin Invest 92(4):1866–1874. https://doi.org/10.1172/JCI116778
99. Mazzone M, Dettori D, de Oliveira RL, Loges S, Schmidt T, Jonckx B, Tian YM, Lanahan AA, Pollard P, de Almodovar CR, De Smet F, Vinckier S, Aragones J, Debackere K, Luttun A, Wyns S, Jordan B, Pisacane A, Gallez B, Lampugnani MG, Dejana E, Simons M, Ratcliffe P, Maxwell P, Carmeliet P (2009) Heterozygous deficiency of PHD2 restores tumor oxygenation and inhibits metastasis via endothelial normalization. Cell 136(5):839–851. https://doi.org/10.1016/j.cell.2009.01.020
100. Meng S, Tripathy D, Frenkel EP, Shete S, Naftalis EZ, Huth JF, Beitsch PD, Leitch M, Hoover S, Euhus D, Haley B, Morrison L, Fleming TP, Herlyn D, Terstappen LW, Fehm T, Tucker TF, Lane N, Wang J, Uhr JW (2004) Circulating tumor cells in patients with breast cancer dormancy. Clin Cancer Res

10(24):8152–8162. https://doi.org/10.1158/1078-0432.CCR-04-1110
101. Miles D, Cameron D, Bondarenko I, Manzyuk L, Alcedo JC, Lopez RI, Im SA, Canon JL, Shparyk Y, Yardley DA, Masuda N, Ro J, Denduluri N, Hubeaux S, Quah C, Bais C, O'Shaughnessy J (2017a) Bevacizumab plus paclitaxel versus placebo plus paclitaxel as first-line therapy for HER2-negative metastatic breast cancer (MERiDiAN): A double-blind placebo-controlled randomised phase III trial with prospective biomarker evaluation. Eur J Cancer 70:146–155. https://doi.org/10.1016/j.ejca.2016.09.024
102. Miles D, Cameron D, Hilton M, Garcia J, O'Shaughnessy J (2017b) Overall survival in MERiDiAN, a double-blind placebo-controlled randomised phase III trial evaluating first-line bevacizumab plus paclitaxel for HER2-negative metastatic breast cancer. Eur J Cancer 90:153–155. https://doi.org/10.1016/j.ejca.2017.10.018
103. Miles DW, Chan A, Dirix LY, Cortes J, Pivot X, Tomczak P, Delozier T, Sohn JH, Provencher L, Puglisi F, Harbeck N, Steger GG, Schneeweiss A, Wardley AM, Chlistalla A, Romieu G (2010) Phase III study of bevacizumab plus docetaxel compared with placebo plus docetaxel for the first-line treatment of human epidermal growth factor receptor 2-negative metastatic breast cancer. J Clin Oncol 28(20):3239–3247. https://doi.org/10.1200/JCO.2008.21.6457
104. Miller K, Wang M, Gralow J, Dickler M, Cobleigh M, Perez EA, Shenkier T, Cella D, Davidson NE (2007) Paclitaxel plus bevacizumab versus paclitaxel alone for metastatic breast cancer. N Engl J Med 357(26):2666–2676. https://doi.org/10.1056/NEJMoa072113
105. Miller KD, Chap LI, Holmes FA, Cobleigh MA, Marcom PK, Fehrenbacher L, Dickler M, Overmoyer BA, Reimann JD, Sing AP, Langmuir V, Rugo HS (2005) Randomized phase III trial of capecitabine compared with bevacizumab plus capecitabine in patients with previously treated metastatic breast cancer. J Clin Oncol 23(4):792–799. https://doi.org/10.1200/JCO.2005.05.098
106. Mittal K, Williams A, Rawal S, Venur VA, Lu B, Torres-Munoz J, Borden EC, Ao Z, OMalley M, Wood LS, Zheng S, Datar RH, Tai YC, Cote RJ, Rini BI (2014) Circulating tumor cell kinetics in mRCC patients treated with sunitinib. J Clin Oncol 32(4). https://doi.org/10.1200/jco.2014.32.4_suppl.481
107. Mountzios G, Pentheroudakis G, Carmeliet P (2014) Bevacizumab and micrometastases: revisiting the preclinical and clinical rollercoaster. Pharmacol Ther 141(2):117–124. https://doi.org/10.1016/j.pharmthera.2013.09.003
108. Nahleh ZA, Barlow WE, Hayes DF, Schott AF, Gralow JR, Sikov WM, Perez EA, Chennuru S, Mirshahidi HR, Corso SW, Lew DL, Pusztai L, Livingston RB, Hortobagyi GN (2016) SWOG S0800 (NCI CDR0000636131): addition of bevacizumab to neoadjuvant nab-paclitaxel with dose-dense doxorubicin and cyclophosphamide improves pathologic complete response (pCR) rates in inflammatory or locally advanced breast cancer. Breast Cancer Res Treat 158(3):485–495. https://doi.org/10.1007/s10549-016-3889-6
109. Naume B, Synnestvedt M, Falk RS, Wiedswang G, Weyde K, Risberg T, Kersten C, Mjaaland I, Vindi L, Sommer HH, Satersdal AB, Rypdal MC, Bendigtsen Schirmer C, Wist EA, Borgen E (2014) Clinical outcome with correlation to disseminated tumor cell (DTC) status after DTC-guided secondary adjuvant treatment with docetaxel in early breast cancer. J Clin Oncol 32(34):3848–3857
110. Paez-Ribes M, Allen E, Hudock J, Takeda T, Okuyama H, Vinals F, Inoue M, Bergers G, Hanahan D, Casanovas O (2009) Antiangiogenic therapy elicits malignant progression of tumors to increased local invasion and distant metastasis. Cancer Cell 15(3):220–231. https://doi.org/10.1016/j.ccr.2009.01.027
111. Paez-Ribes M, Man S, Xu P, Kerbel RS (2015) Potential proinvasive or metastatic effects of preclinical antiangiogenic therapy are prevented by concurrent chemotherapy. Clin Cancer Res 21(24):5488–5498. https://doi.org/10.1158/1078-0432.CCR-15-0915
112. Pantel K, Schlimok G, Braun S, Kutter D, Lindemann F, Schaller G, Funke I, Izbicki JR, Riethmuller G (1993) Differential expression of proliferation-associated molecules in individual micrometastatic carcinoma cells. J Natl Cancer Inst 85(17):1419–1424
113. Park JS, Kim IK, Han S, Park I, Kim C, Bae J, Oh SJ, Lee S, Kim JH, Woo DC, He Y, Augustin HG, Kim I, Lee D, Koh GY (2016) Normalization of tumor vessels by Tie2 activation and Ang2 inhibition enhances drug delivery and produces a favorable tumor microenvironment. Cancer Cell 30(6):953–967. https://doi.org/10.1016/j.ccell.2016.10.018
114. Pennacchietti S, Michieli P, Galluzzo M, Mazzone M, Giordano S, Comoglio PM (2003) Hypoxia promotes invasive growth by transcriptional activation of the met protooncogene. Cancer Cell 3(4):347–361
115. Pezzella F, Di Bacco A, Viale G, Nicholson AG, Price R, Ratcliffe C, Pastorino U, Gatter KC, Altman DG, Harris AL, Pilotti S, Veronesi U (2000) Evidence for novel non-angiogenic pathway in breast-cancer metastasis. Breast cancer progression working party. Lancet 355(9217):1787–1788
116. Phillips RJ, Burdick MD, Lutz M, Belperio JA, Keane MP, Strieter RM (2003) The stromal derived factor-1/CXCL12-CXC chemokine receptor 4 biological axis in non-small cell lung cancer metastases. Am J Respir Crit Care Med 167(12):1676–1686. https://doi.org/10.1164/rccm.200301-071OC
117. Pisarsky L, Bill R, Fagiani E, Dimeloe S, Goosen RW, Hagmann J, Hess C, Christofori G (2016) Targeting metabolic symbiosis to overcome resistance to anti-

angiogenic therapy. Cell Rep 15(6):1161–1174. https://doi.org/10.1016/j.celrep.2016.04.028
118. Pitt LA, Tikhonova AN, Hu H, Trimarchi T, King B, Gong Y, Sanchez-Martin M, Tsirigos A, Littman DR, Ferrando AA, Morrison SJ, Fooksman DR, Aifantis I, Schwab SR (2015) CXCL12-producing vascular endothelial niches control acute T cell leukemia maintenance. Cancer Cell 27(6):755–768. https://doi.org/10.1016/j.ccell.2015.05.002
119. Pivot X, Schneeweiss A, Verma S, Thomssen C, Passos-Coelho JL, Benedetti G, Ciruelos E, von Moos R, Chang HT, Duenne AA, Miles DW (2011) Efficacy and safety of bevacizumab in combination with docetaxel for the first-line treatment of elderly patients with locally recurrent or metastatic breast cancer: results from AVADO. Eur J Cancer 47(16):2387–2395. https://doi.org/10.1016/j.ejca.2011.06.018
120. Price TT, Burness ML, Sivan A, Warner MJ, Cheng R, Lee CH, Olivere L, Comatas K, Magnani J, Kim Lyerly H, Cheng Q, McCall CM, Sipkins DA (2016) Dormant breast cancer micrometastases reside in specific bone marrow niches that regulate their transit to and from bone. Sci Transl Med 8(340):340ra373. https://doi.org/10.1126/scitranslmed.aad4059
121. Provenzano PP, Cuevas C, Chang AE, Goel VK, Von Hoff DD, Hingorani SR (2012) Enzymatic targeting of the stroma ablates physical barriers to treatment of pancreatic ductal adenocarcinoma. Cancer Cell 21(3):418–429. https://doi.org/10.1016/j.ccr.2012.01.007
122. Qian BZ, Li J, Zhang H, Kitamura T, Zhang J, Campion LR, Kaiser EA, Snyder LA, Pollard JW (2011) CCL2 recruits inflammatory monocytes to facilitate breast-tumour metastasis. Nature 475(7355):222–225. https://doi.org/10.1038/nature10138
123. Rafii S, Butler JM, Ding BS (2016) Angiocrine functions of organ-specific endothelial cells. Nature 529(7586):316–325. https://doi.org/10.1038/nature17040
124. Rajendran P, Rengarajan T, Thangavel J, Nishigaki Y, Sakthisekaran D, Sethi G, Nishigaki I (2013) The vascular endothelium and human diseases. Int J Biol Sci 9(10):1057–1069. https://doi.org/10.7150/ijbs.7502
125. Rakhra K, Bachireddy P, Zabuawala T, Zeiser R, Xu L, Kopelman A, Fan AC, Yang Q, Braunstein L, Crosby E, Ryeom S, Felsher DW (2010) CD4(+) T cells contribute to the remodeling of the microenvironment required for sustained tumor regression upon oncogene inactivation. Cancer Cell 18(5):485–498. https://doi.org/10.1016/j.ccr.2010.10.002
126. Ravaud A, Motzer RJ, Pandha HS, George DJ, Pantuck AJ, Patel A, Chang YH, Escudier B, Donskov F, Magheli A, Carteni G, Laguerre B, Tomczak P, Breza J, Gerletti P, Lechuga M, Lin X, Martini JF, Ramaswamy K, Casey M, Staehler M, Patard JJ, ST Investigators (2016) Adjuvant sunitinib in high-risk renal-cell carcinoma after nephrectomy. N Engl J Med 375(23):2246–2254. https://doi.org/10.1056/NEJMoa1611406
127. Raymond E, Dahan L, Raoul JL, Bang YJ, Borbath I, Lombard-Bohas C, Valle J, Metrakos P, Smith D, Vinik A, Chen JS, Horsch D, Hammel P, Wiedenmann B, Van Cutsem E, Patyna S, Lu DR, Blanckmeister C, Chao R, Ruszniewski P (2011) Sunitinib malate for the treatment of pancreatic neuroendocrine tumors. N Engl J Med 364(6):501–513. https://doi.org/10.1056/NEJMoa1003825
128. Relf M, LeJeune S, Scott PA, Fox S, Smith K, Leek R, Moghaddam A, Whitehouse R, Bicknell R, Harris AL (1997) Expression of the angiogenic factors vascular endothelial cell growth factor, acidic and basic fibroblast growth factor, tumor growth factor beta-1, platelet-derived endothelial cell growth factor, placenta growth factor, and pleiotrophin in human primary breast cancer and its relation to angiogenesis. Cancer Res 57(5):963–969
129. Robert NJ, Dieras V, Glaspy J, Brufsky A, Bondarenko I, Lipatov O, Perez E, Yardley D, Zhou X, Phan S (2009) RIBBON-1: Randomized, double-blind, placebo-controlled, phase III trial of chemotherapy with or without bevacizumab (B) for first-line treatment of HER2-negative locally recurrent or metastatic breast cancer (MBC). J Clin Oncol 27(15_suppl):1005
130. Robert NJ, Dieras V, Glaspy J, Brufsky AM, Bondarenko I, Lipatov ON, Perez EA, Yardley DA, Chan SYT, Zhou X, Phan S-C, O'Shaughnessy J (2011a) RIBBON-1: randomized, double-blind, placebo-controlled, phase III trial of chemotherapy with or without bevacizumab for first-line treatment of human epidermal growth factor receptor 2-negative, locally recurrent or metastatic breast cancer. J Clin Oncol 29(10):1252–1260
131. Robert NJ, Saleh MN, Paul D, Generali D, Gressot L, Copur MS, Brufsky AM, Minton SE, Giguere JK, Smith JW 2nd, Richards PD, Gernhardt D, Huang X, Liau KF, Kern KA, Davis J (2011b) Sunitinib plus paclitaxel versus bevacizumab plus paclitaxel for first-line treatment of patients with advanced breast cancer: a phase III, randomized, open-label trial. Clin Breast Cancer 11(2):82–92. https://doi.org/10.1016/j.clbc.2011.03.005
132. Russell WE, McGowan JA, Bucher NL (1984) Partial characterization of a hepatocyte growth factor from rat platelets. J Cell Physiol 119(2):183–192. https://doi.org/10.1002/jcp.1041190207
133. Sanger N, Effenberger KE, Riethdorf S, Van Haasteren V, Gauwerky J, Wiegratz I, Strebhardt K, Kaufmann M, Pantel K (2011) Disseminated tumor cells in the bone marrow of patients with ductal carcinoma in situ. Int J Cancer 129(10):2522–2526. https://doi.org/10.1002/ijc.25895
134. Schwartzberg LS, Tauer KW, Hermann RC, Makari-Judson G, Isaacs C, Beck JT, Kaklamani V, Stepanski EJ, Rugo HS, Wang W, Bell-McGuinn K, Kirshner JJ, Eisenberg P, Emanuelson R, Keaton M, Levine E, Medgyesy DC, Qamar R, Starr A,

Ro SK, Lokker NA, Hudis CA (2013) Sorafenib or placebo with either gemcitabine or capecitabine in patients with HER-2-negative advanced breast cancer that progressed during or after bevacizumab. Clin Cancer Res 19(10):2745–2754. https://doi.org/10.1158/1078-0432.CCR-12-3177
135. Seaman S, Zhu Z, Saha S, Zhang XM, Yang MY, Hilton MB, Morris K, Szot C, Morris H, Swing DA, Tessarollo L, Smith SW, Degrado S, Borkin D, Jain N, Scheiermann J, Feng Y, Wang Y, Li J, Welsch D, DeCrescenzo G, Chaudhary A, Zudaire E, Klarmann KD, Keller JR, Dimitrov DS, St Croix B (2017) Eradication of tumors through simultaneous ablation of CD276/B7-H3-positive tumor cells and tumor vasculature. Cancer Cell 31(4):501–515 e508. https://doi.org/10.1016/j.ccell.2017.03.005
136. Sennino B, Ishiguro-Oonuma T, Wei Y, Naylor RM, Williamson CW, Bhagwandin V, Tabruyn SP, You WK, Chapman HA, Christensen JG, Aftab DT, McDonald DM (2012) Suppression of tumor invasion and metastasis by concurrent inhibition of c-Met and VEGF signaling in pancreatic neuroendocrine tumors. Cancer Discov 2(3):270–287. https://doi.org/10.1158/2159-8290.CD-11-0240
137. Shojaei F, Wu X, Malik AK, Zhong C, Baldwin ME, Schanz S, Fuh G, Gerber HP, Ferrara N (2007a) Tumor refractoriness to anti-VEGF treatment is mediated by CD11b+Gr1+ myeloid cells. Nat Biotechnol 25(8):911–920. https://doi.org/10.1038/nbt1323
138. Shojaei F, Wu X, Zhong C, Yu L, Liang XH, Yao J, Blanchard D, Bais C, Peale FV, van Bruggen N, Ho C, Ross J, Tan M, Carano RA, Meng YG, Ferrara N (2007b) Bv8 regulates myeloid-cell-dependent tumour angiogenesis. Nature 450(7171):825–831. https://doi.org/10.1038/nature06348
139. Sikov WM, Berry DA, Perou CM, Singh B, Cirrincione CT, Tolaney SM, Kuzma CS, Pluard TJ, Somlo G, Port ER, Golshan M, Bellon JR, Collyar D, Hahn OM, Carey LA, Hudis CA, Winer EP (2015) Impact of the addition of carboplatin and/or bevacizumab to neoadjuvant once-per-week paclitaxel followed by dose-dense doxorubicin and cyclophosphamide on pathologic complete response rates in stage II to III triple-negative breast cancer: CALGB 40603 (Alliance). J Clin Oncol 33(1):13–21. https://doi.org/10.1200/JCO.2014.57.0572
140. Sosa MS, Bragado P, Aguirre-Ghiso JA (2014) Mechanisms of disseminated cancer cell dormancy: an awakening field. Nat Rev Cancer 14(9):611–622. https://doi.org/10.1038/nrc3793
141. Sounni NE, Cimino J, Blacher S, Primac I, Truong A, Mazzucchelli G, Paye A, Calligaris D, Debois D, De Tullio P, Mari B, De Pauw E, Noel A (2014) Blocking lipid synthesis overcomes tumor regrowth and metastasis after antiangiogenic therapy withdrawal. Cell Metab 20(2):280–294. https://doi.org/10.1016/j.cmet.2014.05.022
142. Spiliotaki M, Mavroudis D, Kapranou K, Markomanolaki H, Kallergi G, Koinis F, Kalbakis K, Georgoulias V, Agelaki S (2014) Evaluation of proliferation and apoptosis markers in circulating tumor cells of women with early breast cancer who are candidates for tumor dormancy. Breast Cancer Res 16(6):485. https://doi.org/10.1186/s13058-014-0485-8
143. Srivastava K, Hu J, Korn C, Savant S, Teichert M, Kapel SS, Jugold M, Besemfelder E, Thomas M, Pasparakis M, Augustin HG (2014) Postsurgical adjuvant tumor therapy by combining anti-angiopoietin-2 and metronomic chemotherapy limits metastatic growth. Cancer Cell 26(6):880–895. https://doi.org/10.1016/j.ccell.2014.11.005
144. Stawowczyk M, Wellenstein MD, Lee SB, Yomtoubian S, Durrans A, Choi H, Narula N, Altorki NK, Gao D, Mittal V (2017) Matrix Metalloproteinase 14 promotes lung cancer by cleavage of Heparin-Binding EGF-like Growth Factor. Neoplasia (New York, N Y) 19(2):55–64
145. Sweeney CJ, Chiorean EG, Verschraegen CF, Lee FC, Jones S, Royce M, Tye L, Liau KF, Bello A, Chao R, Burris HA (2010) A phase I study of sunitinib plus capecitabine in patients with advanced solid tumors. J Clin Oncol 28(29):4513–4520. https://doi.org/10.1200/JCO.2009.26.9696
146. Thorban S, Roder JD, Nekarda H, Funk A, Pantel K, Siewert JR (1996) Disseminated epithelial tumor cells in bone marrow of patients with esophageal cancer: detection and prognostic significance. World J Surg 20(5):567–572; discussion 572–563
147. Tian L, Goldstein A, Wang H, Ching Lo H, Sun Kim I, Welte T, Sheng K, Dobrolecki LE, Zhang X, Putluri N, Phung TL, Mani SA, Stossi F, Sreekumar A, Mancini MA, Decker WK, Zong C, Lewis MT, Zhang XH (2017) Mutual regulation of tumour vessel normalization and immunostimulatory reprogramming. Nature 544(7649):250–254. https://doi.org/10.1038/nature21724
148. van der Bilt JD, Kranenburg O, Nijkamp MW, Smakman N, Veenendaal LM, Te Velde EA, Voest EE, van Diest PJ, Borel Rinkes IH (2005) Ischemia/reperfusion accelerates the outgrowth of hepatic micrometastases in a highly standardized murine model. Hepatology 42(1):165–175. https://doi.org/10.1002/hep.20739
149. van der Veldt AA, Vroling L, de Haas RR, Koolwijk P, van den Eertwegh AJ, Haanen JB, van Hinsbergh VW, Broxterman HJ, Boven E (2012) Sunitinib-induced changes in circulating endothelial cell-related proteins in patients with metastatic renal cell cancer. Int J Cancer 131(4):E484–E493. https://doi.org/10.1002/ijc.26456
150. Vessella RL, Pantel K, Mohla S (2007) Tumor cell dormancy: an NCI workshop report. Cancer Biol Ther 6(9):1496–1504
151. von Minckwitz G, Loibl S, Untch M, Eidtmann H, Rezai M, Fasching PA, Tesch H, Eggemann H, Schrader I, Kittel K, Hanusch C, Huober J, Solbach C, Jackisch C, Kunz G, Blohmer JU, Hauschild M, Fehm T, Nekljudova V, Gerber B, GA-Bs groups

(2014a) Survival after neoadjuvant chemotherapy with or without bevacizumab or everolimus for HER2-negative primary breast cancer (GBG 44-GeparQuinto)dagger. Ann Oncol 25(12):2363–2372. https://doi.org/10.1093/annonc/mdu455
152. von Minckwitz G, Puglisi F, Cortes J, Vrdoljak E, Marschner N, Zielinski C, Villanueva C, Romieu G, Lang I, Ciruelos E, De Laurentiis M, Veyret C, de Ducla S, Freudensprung U, Srock S, Gligorov J (2014b) Bevacizumab plus chemotherapy versus chemotherapy alone as second-line treatment for patients with HER2-negative locally recurrent or metastatic breast cancer after first-line treatment with bevacizumab plus chemotherapy (TANIA): an open-label, randomised phase 3 trial. The Lancet Oncol 15(11):1269–1278. https://doi.org/10.1016/S1470-2045(14)70439-5
153. Vrdoljak E, Marschner N, Zielinski C, Gligorov J, Cortes J, Puglisi F, Aapro M, Fallowfield L, Fontana A, Inbar M, Kahan Z, Welt A, Levy C, Brain E, Pivot X, Putzu C, Gonzalez Martin A, de Ducla S, Easton V, von Minckwitz G (2016) Final results of the TANIA randomised phase III trial of bevacizumab after progression on first-line bevacizumab therapy for HER2-negative locally recurrent/metastatic breast cancer. Ann Oncol 27(11):2046–2052. https://doi.org/10.1093/annonc/mdw316
154. Wang H, Fu W, Im JH, Zhou Z, Santoro SA, Iyer V, DiPersio CM, Yu QC, Quaranta V, Al-Mehdi A, Muschel RJ (2004) Tumor cell alpha3beta1 integrin and vascular laminin-5 mediate pulmonary arrest and metastasis. J Cell Biol 164(6):935–941. https://doi.org/10.1083/jcb.200309112
155. Weis S, Cui J, Barnes L, Cheresh D (2004) Endothelial barrier disruption by VEGF-mediated Src activity potentiates tumor cell extravasation and metastasis. J Cell Biol 167(2):223–229. https://doi.org/10.1083/jcb.200408130
156. Wertheimer T, Velardi E, Tsai J, Cooper K, Xiao S, Kloss CC, Ottmuller KJ, Mokhtari Z, Brede C, deRoos P, Kinsella S, Palikuqi B, Ginsberg M, Young LF, Kreines F, Lieberman SR, Lazrak A, Guo P, Malard F, Smith OM, Shono Y, Jenq RR, Hanash AM, Nolan DJ, Butler JM, Beilhack A, Manley NR, Rafii S, Dudakov JA, van den Brink MRM (2018) Production of BMP4 by endothelial cells is crucial for endogenous thymic regeneration. Sci Immunol 3(19). https://doi.org/10.1126/sciimmunol.aal2736
157. Wiedswang G, Borgen E, Karesen R, Kvalheim G, Nesland JM, Qvist H, Schlichting E, Sauer T, Janbu J, Harbitz T, Naume B (2003) Detection of isolated tumor cells in bone marrow is an independent prognostic factor in breast cancer. J Clin Oncol 21(18):3469–3478. https://doi.org/10.1200/JCO.2003.02.009
158. Wong AL, Sundar R, Wang TT, Ng TC, Zhang B, Tan SH, Soh TI, Pang AS, Tan CS, Ow SG, Wang L, Mogro J, Ho J, Jeyasekharan AD, Huang Y, Thng CH, Chan CW, Hartman M, Iau P, Buhari SA, Goh BC, Lee SC (2016) Phase Ib/II randomized, open-label study of doxorubicin and cyclophosphamide with or without low-dose, short-course sunitinib in the pre-operative treatment of breast cancer. Oncotarget 7(39):64089–64099. https://doi.org/10.18632/oncotarget.11596
159. Xian X, Hakansson J, Stahlberg A, Lindblom P, Betsholtz C, Gerhardt H, Semb H (2006) Pericytes limit tumor cell metastasis. J Clin Invest 116(3):642–651. https://doi.org/10.1172/JCI25705
160. Yang Y, Zhang Y, Iwamoto H, Hosaka K, Seki T, Andersson P, Lim S, Fischer C, Nakamura M, Abe M, Cao R, Skov PV, Chen F, Chen X, Lu Y, Nie G, Cao Y (2016) Discontinuation of anti-VEGF cancer therapy promotes metastasis through a liver revascularization mechanism. Nat Commun 7:12680. https://doi.org/10.1038/ncomms12680
161. Yumoto K, Eber MR, Wang J, Cackowski FC, Decker AM, Lee E, Nobre AR, Aguirre-Ghiso JA, Jung Y, Taichman RS (2016) Axl is required for TGF-beta2-induced dormancy of prostate cancer cells in the bone marrow. Sci Rep 6:36520. https://doi.org/10.1038/srep36520
162. Zhu TS, Costello MA, Talsma CE, Flack CG, Crowley JG, Hamm LL, He X, Hervey-Jumper SL, Heth JA, Muraszko KM, DiMeco F, Vescovi AL, Fan X (2011) Endothelial cells create a stem cell niche in glioblastoma by providing NOTCH ligands that nurture self-renewal of cancer stem-like cells. Cancer Res 71(18):6061–6072. https://doi.org/10.1158/0008-5472.CAN-10-4269

Minimal Residual Disease in Prostate Cancer

3

Frank C. Cackowski and Russell S. Taichman

Abstract

Detection of minimal residual disease (MRD) in prostate cancer over several decades has greatly informed our understanding of dissemination and recurrence, but has not yet been routinely used in clinical care. Investigators have detected MRD by identification of prostate cancer cells in the bone marrow; termed disseminated tumor cells (DTCs) and blood; termed circulating tumor cells (CTCs). Various techniques including PSA-RT PCR, PSA immunocytochemistry, cytokeratin immunocytochemistry, and immune-magnetic depletion of hematopoietic cells followed by EPCAM based positive selection, have been used. Importantly, detection of DTCs correlates with recurrence. Research into prostate cancer CTCs has intensified recently, but their use in MRD evaluation has been more limited. Investigators are using semi-automated platforms to detect and begin to study prostate cancer CTCs in patients with no evidence of disease. PSA immunocytochemistry also detects CTCs and correlates with recurrence. Emerging technologies have the potential to greatly aid research in this exciting field.

Keywords

Prostate cancer · Disseminated tumor cell · Circulating tumor cell · Dormancy · PSA-RT PCR · EPCAM · CTC · Liquid biopsy · Bone marrow · Bone metastasis

3.1 Introduction

Prostate cancer is a common and serious clinical problem as it causes over 26,000 deaths per year in the United States [1]. Although progress has been made, like most epithelial cancers, metastatic prostate cancer remains incurable. However, unlike most epithelial cancers, clinicians treating prostate cancer have had the blessing (or curse) of a sensitive and specific blood tumor marker, the prostate specific antigen (PSA) for over 30 years [2]. Although it's utility as a screening test for new prostate cancer is debated, PSA is quite good at detecting low levels of prostate cancer (or normal prostate tissue) in patients who have had a localized prostate cancer treated

F. C. Cackowski
Department of Periodontics and Oral Medicine, University of Michigan School of Dentistry, Ann Arbor, MI, USA

Department of Medicine, Division of Hematology & Oncology, University of Michigan School of Medicine, Ann Arbor, MI, USA

R. S. Taichman (✉)
Department of Periodontics and Oral Medicine, University of Michigan School of Dentistry, Ann Arbor, MI, USA
e-mail: rtaich@umich.edu

J. A. Aguirre-Ghiso (ed.), *Biological Mechanisms of Minimal Residual Disease and Systemic Cancer*, Advances in Experimental Medicine and Biology 1100,
https://doi.org/10.1007/978-3-319-97746-1_3

by surgery or radiation. Therefore, PSA is nearly universally used to monitor for prostate cancer recurrence and to monitor treatment of advanced disease if recurrence does occur. However, the fact that patients can have an undetectable PSA after surgery and subsequently recur illustrates that PSA measurement is not capable of detecting the lowest levels of clinically significant disease [3]. Therefore, investigators have endeavored to detect lower levels of disease, usually by the detection of intact prostate tumor cells.

When detected in peripheral blood, tumor cells are often termed circulating tumor cells (CTCs). Because of the fact that prostate cancer metastasizes commonly to bone, and presumably because of tissue accessibility, investigators have also reported prostate cancer cells in bone marrow when there is no disease detected elsewhere. In this circumstance, the cells are usually termed disseminated tumor cells (DTCs). Although DTCs may also be present in other tissues, little work has been performed with patient samples in an attempt to identify prostate cancer cells in tissues that can also host metastatic disease such as lymph nodes, lung and liver. Yet unlike some hematologic cancers, prostate cancer minimal residual disease detection has not yet received regular clinical use. However, these techniques have been invaluable for understanding the time-course of prostate cancer dissemination, and generating hypotheses for subsequent testing in model systems or conversely validating observations derived from model systems.

Here, we discuss the use of CTC and DTC detection to identify patients with residual disease who would have been presumed to be free of disease by PSA and imaging criteria. We also include studies of patients with clinically localized prostate cancer, but detectable bone marrow DTCs at the time of radical prostatectomy as these studies impact the risk of recurrence and help explain the biology of the disease. Advances have also been made recently using advanced imaging techniques such pelvic MRI, choline-PET scans and PET scans radio-labeled ligands for prostate specific membrane antigen [4, 5]. As these techniques are usually applied to patients with a detectable PSA, we will not discuss them here as they are not capable at present to detect as low a level of disease as CTC or DTC detection. Despite the fact that techniques for CTC or DTC detection have aided our understanding of prostate cancer biology, they are still limited in terms of reproducibility between research groups, purity of the isolated cells and the information that can be gained after isolation. We hope to compile and describe the work performed to help our readers continue the progress in this essential avenue of research.

3.2 Bone Marrow Disseminated Tumor Cells

It is now over 20 years since prostate cancer bone marrow DTCs were reported in patients with no evidence of disease outside of the prostate [6]. The presence of these cells along with observation that about 20% of prostate cancer recurrences occur greater than 5 years after curative intent surgery has provided the clinical basis for the field of research into cancer dormancy and dormancy escape. These discoveries in prostate cancer followed identification of bone marrow DTCs in clinically localized breast cancer [7]. This also occurred at a time when our understanding, through the work of Bernard Fisher and others, that cancer dissemination often occurred early in the disease process and therefore that the disfiguring radical mastectomy surgeries were not required, especially if more conservative surgery was coupled with adjuvant radiation or systemic therapy. There was also a report [8] in the 1980s of prostate cancer cells detected in bone marrow patients with localized disease by immunohistochemistry and occasionally similar reports for decades before this – reviewed in [9]. The reports discussed below of the detection of prostate cancer DTCs in patients with localized disease or otherwise undetectable disease are summarized in Table 3.1.

The work in this area became accelerated in the 1990s with the use of the then relatively new technique of reverse transcriptase polymerase chain reaction (RT-PCR) for PSA (*KLK3*). Wood

Table 3.1 Summary of the literature describing detection of bone marrow disseminated tumor cells (DTCs) in prostate cancer patients with clinically localized disease or otherwise undetectable disease

Technique:	DTCs detected?	Correlated with recurrence?	Reference (s)
PSA RT-PCR	Yes	Yes	[6, 9, 10, 16]
Cytokeratin immunocytochemistry	Yes	Yes	[7–9, 12–14]
Immunomagnentic negative and positive ± selection single EPCAM positive cell isolation	Yes	Yes	[15–19]
PSA immunocytochemistry	Yes	Yes	[21, 22, 26]
Various: AdnaTest, VERSA, HD-SCA, RareCyte	No/few	Not applicable	[24]

et al. [6] combined Southern hybridization in the early 1990s to visualize the low abundance PSA RT-PCR product in bone marrow aspirates from patients with benign prostate hyperplasia (as negative controls), localized or metastatic prostate cancer. The presence or absence of distant metastases was determined by bone scan. PSA mRNA was detected in all 5 of the metastatic patients and none of the five negative controls. PSA was detected in 19 of 43 (44%) of localized prostate cancer patients with negative bone scans and negative lymph nodes by histology, and in 5 of 7 patients with involved lymph nodes but no macroscopic bone disease. The also performed immunohistochemistry (IHC) for PSA on bone marrow aspirates that had PSA detected by RT-PCR and detected PSA positive cells in 24 of 29 patients for whom samples were available. The frequency of PSA positive cells ranged between 1 and 30 per million bone marrow cells. Some of these same authors later reported clinical follow-up of 86 patients with localized prostate cancer treated with radical prostatectomy and a bone marrow aspirate collected between 4 and 8 weeks afterwards and assayed by RT-PCR for PSA [10]. Importantly the presence of bone marrow DTCs (PSA positivity) predicted a worse disease free survival – which supports the validity of this approach and the importance of bone marrow DTCs. Also in the 1990s, Melchior et al., also used RT-PCR to detect DTCs in the bone marrow of 71 patients with localized prostate cancer [9]. In this report none of the patients without prostate cancer were RT-PCR PSA positive, whereas 56% of patients with organ confined (pT2) prostate cancer and 73% of patients with extra-capsular extension had PSA positive by RT-PCR from bone marrow. They also assayed paired peripheral blood samples and reported lower rates of detection. Similarly, they found lower sensitivity with immunocytochemistry. Overall, the work in the 1990s using RT-PCR for PSA showed that prostate cancer cells disseminated to the bone marrow early in the disease process and because of the correlation with clinical outcomes, suggested that they might predict disease relapse. However, a major drawback of these studies was that they did not allow subsequent isolation of the detected cells for subsequent analysis.

In parallel with this work, and in parallel with studies in breast cancer [11], investigators in Germany also detected putative DTCs in the bone marrow of localized prostate cancer patients with independent techniques – immunocytochemistry for cytokeratins. In 1995, Pantel and colleagues reported cells positive for epithelial cytokeratin 18 in 24 out of 44 patients with stage C (modern stage T3a or T3b) localized prostate cancer [12]. They performed these analyses on bilateral bone marrow aspirates, 10^6 mononuclear cells per side, and detected between one and 38 cytokeratin 18 positive cells per patient. The authors also noted that cytokeratin positive cells were only noted in 13 of the 24 positive patients. The authors do not report results from bone marrow of normal controls and did not observe a correlation between the presence of DTCs and serum PSA or histologic grade and did not have an adequate length of follow-up to report any correlation with cancer recurrence. Subsequently, authors from the same group analyzed bone marrow aspirates from 82 patients prior to radical prostatectomy with a pan-cytokeratin antibody (epitope common to epitopes from cytokeratin 8, 18 and 19) or an antibody against cytokeratin 18 [13]. The patients with pan-cytokeratin positive DTCs in the bone marrow had a biochemical relapse free survival of 886 days, versus 1409 days in pan-cytokeratin negative patients. Furthermore, pan-cytokeratin positivity remained an independent predictor of recurrence when controlled for PSA concentration, pathological stage, and Gleason score. Conversely, they did not observe correlation with clinical parameters in cells identified with the antibody against cytokeratin 18 and hypothesized that the pan-cytokeratin antibody might have resulted in less false negatives. Finally, some of the same investigators reported the presence of pan-cytokeratin positive DTCs in 86 of 193 (44.6%) clinical lymph node negative localized prostate cancer patients prior to neoadjuvant androgen deprivation and subsequent radical prostatectomy (not currently a standard therapeutic approach) [14]. Similar to the prior report, they also found that presence of DTCs was an independent predictor of biochemical recurrence.

Also in the 2000s investigators, primarily in Dr. Robert Vessella's group at the University of Washington, began to develop techniques that would eventually allow the isolation of DTCs for further study. In 2003, Ellis and colleagues reported techniques using immuno-magnetic methods to enrich for epithelial cells in the bone marrow presumed to be prostate DTCs. They performed negative selection for hematopoietic cell markers CD45 and CD61, then positive selection for human epithelial cell antigen (EPCAM) [15]. The isolated cells were subsequently subjected to RT-PCR for PSA. Of 126 patients with localized prostate cancer, 54% had PSA positive epithelial cells before prostatectomy and 33% had PSA positive epithelial cells a median of 4 months after prostatectomy. It is important to note that RT-PCR for PSA was positive in only 60% of patients with bone marrow epithelial cells. Presuming that PSA is a good marker in this circumstance, this suggests that EPCAM is not completely specific. One of ten negative controls had bone marrow epithelial cells, but this sample was PSA negative. The same group also reported higher rates of DTC detection using this immunomagnetic based enrichment as compared to direct RT-PCR of bone marrow. They also reported the ability to capture these cells with a micromanipulator for subsequent analyses using EPCAM immunofluorescence of viable cells after prior immune-magnetic negative and positive selection [16].

Using these techniques, it was possible to analyze DTCs collected from either localized or metastatic prostate cancer patients for large scale genomic changes using comparative genomic hybridization [17]. In this work, Holcomb et al. reported concordant regions of genomic gain or loss between DTCs isolated from localized patients and their primary tumors. Among these alterations were changes characteristic of prostate cancer including 8p loss, 8q gain, and Xq gain which includes the androgen receptor gene. They reported fewer alterations in DTCs from localized patients than from metastatic patients. The reason for this is unclear, though they argued against dilution from normal cells as the cause. As further evidence that the isolated cells really were tumor cells, after two additional years of follow-up the same group reported correlations between the presence of bone marrow DTCs with shorter time to relapse [18]. In this work, the data was based on live cell immunofluorescent detection of EPCAM after prior immuno-magnetic positive and negative selection. They reported as positive the presence of one DTC in from a tube of bone marrow aspirate, which contains about 10^7 nucleated cells. Using this criteria, DTCs were present in 408 of 569 (72%) of localized prostate cancer patients prior to radical prostatectomy and 3 of 34 (8.8%) negative control patients without prostate cancer. After prostatectomy, DTCs were present in 56 of 98 (57%) of patients with no evidence of disease (undetectable PSA) and in 12 of 14 patients who subsequently suffered biochemical relapse. Presence of DTCs after prostatectomy did not correlate with any one clinicopathologic variable (pre-operative PSA, pathologic stage, Gleason grade, surgical margin, age or race). However, more importantly, the presence of DTCs in patients with no evidence of disease after radical prostatectomy was an independent predictor of biochemical relapse.

In a follow up of their work, the Washington group was later able to perform single cell microarray analyses on DTCs isolated in this fashion from either patients who previously had localized prostate cancer and currently had no evidence of disease or detectable disease [19]. Their data illustrates some of the limitations of an EPCAM based approach for bone marrow DTC isolation – EPCAM expression by normal bone marrow cells. Of the 85 analyzed cells, they excluded 41 cells because their gene expression signature resembled erythroid precursors. However, with the data from the remaining cells, they were able to generate useful gene expression analyses and compile lists of genes differentially regulated in patients with no evidence of disease or active (advanced) disease. This data suggested that p38 MAPK is a critical node controlling the switch from dormancy to proliferation – a finding which had been independently discovered in cell lines and animal models [20].

Others have continued to show the presence of bone marrow DTCs in patients with prostate cancer [21, 22]. In the most recent study, the authors showed that the presence of DTCs but absence of CTCs as detected by immunocytochemistry for PSA correlated with late relapse [22]. This observation coincides with theories of cancer dormancy derived primarily from work in model systems, in which bone marrow DTCs are dormant for long time periods and are a reservoir for subsequent reactivation and late relapse [23].

In contrast with this body of work, a recent report by Chalfin et al., comprising data compiled from investigators at 4 independent institutions detected few DTCs in the bone marrow of patients with localized prostate cancer [24]. The groups used four different proprietary platforms (AdnaTest, high definition single cell assay, VERSA and RareCyte) to analyze bone marrow from 208 patients with localized prostate cancer, 16 control patients without prostate cancer, and 5 patients with metastatic prostate cancer. Using the AdnaTest of gene expression after immune-magnetic selection, 1 of 127 localized and 3 of 5 metastatic prostate cancer patients had PSA detected in bone marrow. Using the VERSA test of gene expression, 1 of 18 patients with localized prostate showed bone marrow PSMA (*FOLH1*) expression. With the high definition single cell assay (HD-SCA), the investigators evaluated androgen receptor expression and localization, and CD45 as a negative marker. Only one of 43 bone marrow samples had a questionable presence of DTCs, though it was below their set threshold number of cells. Using the RareCyte platform for multi-parameter immunofluorescence, the investigators identified one putative DTC in one patient out of 42 patients with localized prostate cancer.

It is difficult to reconcile the data presented by Chalfin et al., with the multiple prior reports of bone marrow DTCs detected in patients with localized prostate cancer. Clearly, DTCs are rare in the marrow early on in disease and possibly at a lower prevalence than previously thought. However, the assays used by Chalfin et al., might lack adequate sensitivity for their detection. DTCs when present are certain to be rare. Other studies reported a frequency of 1 cell in 10^6 or even 10^7 nucleated marrow cells. These investigators reported positive control (metastatic patient) data for the AdnaTest but not the other three platforms. Furthermore, the lack of detection of CTCs with the high definition single cell assay is disconcerting. Further studies are needed to reconcile these discrepancies.

3.3 Circulating Tumor Cells

Research into the use of CTCs, often with semi-automated platforms, in patients with active disease to understand prognosis, predict therapeutic response and probe prostate cancer biology has intensified recently and has been reviewed elsewhere [25]. Here, we focus on the use of CTCs in patients with no evidence of disease such as after radical prostatectomy, termed by some authors as "secondary CTCs [26]." A recent study began to explore the use of three of these platforms to detect CTCs in 107 high risk localized (PSA ≥20 ng/ml, biopsy Gleason score ≥8, or clinical

T stage ≥2c) prostate cancer patients, including 30 patients with paired blood samples drawn before and approximately 3 months after radical prostatectomy [27]. They conducted their analyses with the following systems: CellSearch™ (EPCAM based, FDA cleared), CellCollector™ (EPCAM based *in vivo* collection, and EPISPOT™ (pan-cytokeratin selection followed by assay for PSA secretion). They detected CTCs in 37%, 54.9%, and 58.7% using CellSearch, CellCollector and EPISPOT respectively. With the CellCollector system, they detected a significant decrease from 66% CTC positivity to 34% CTC positivity before and after surgery. Some of the same investigators have subsequently used the same techniques to begin to study the biology of CTCs in patients with minimal residual disease and found that CTCs have more mesenchymal characteristics after radical prostatectomy [28].

Other investigators have also recently used immunocytochemistry for PSA to try and understand the relative significance of DTCs and CTCs in the same patient and whether there might be different types of minimal residual disease [22]. They determined the presence of CTCs and DTCs in 321 patients 1 month after radical prostatectomy and analyzed outcomes in patients that had CTCs, DTCs or both. As expected, recurrence free survival was longest in the 140 patients that had neither CTCs, nor DTCs – a 10 year recurrence free survival of 92.7%. Patients with CTCs present, either with or without DTCs, did poorly with 10 year recurrence free survivals of 5% for the 88 patients with DTCs and 6.4% for the 54 patients without DTCs. The 39 patients with DTCs but not CTCs had an intermediate prognosis, with a 10 year recurrence free survival of 55.8%. The authors also noted that relapses in these patients began to occur more rapidly than expected after about 8 years. They therefore hypothesized that patients with CTCs present had more active disease and that patients without CTCs but with DTCs had more dormant disease and therefore recurred later. These observations are in keeping with laboratory based studies finding that some sites such as the bone marrow cause DTCs to assume a dormant phenotype that is yet still susceptible to late relapse [23].

3.4 Perspective and Future Directions

Analysis of prostate cancer minimal residual disease by detection of CTCs and DTCs over the past several decades has greatly increased our understanding of the biology of dissemination, dormancy and recurrence but has not yet impacted routine clinical care. Because of rapidly advancing technologies, we expect these scientific advances to continue and possibly more directly impact patient care by prognostication or by informing adjuvant therapy. New techniques, especially next generation sequencing based approaches have the potential to greatly impact our ability to detect prostate cancer minimal residual disease and to be more confident in the results of the testing.

Financial Support Direct funding was provided by the NIH/NCI P01-CA093900, Department of Defense W81XWH-14-1-0403, W81XWH-15-1-0413, the Prostate Cancer Foundation Challenge award 16CHAL05. R.T. receives support as the Major McKinley Ash Colligate Professor. F.C. receives support from a Career Enhancement Award from the NIH/NCI Prostate Cancer Specialized Program in Research Excellence (SPORE) #F048931, sub-award of #F036250 at the University of Michigan.

References

1. Zhang J, Guo X, Chang DY, Rosen DG, Mercado-Uribe I, Liu J (2012) CD133 expression associated with poor prognosis in ovarian cancer 3. Mod Pathol 25(3):456–464
2. Kuriyama M, Wang MC, Papsidero LD et al (1980) Quantitation of prostate-specific antigen in serum by a sensitive enzyme immunoassay. Cancer Res 40(12):4658–4662
3. Pound CR, Partin AW, Eisenberger MA, Chan DW, Pearson JD, Walsh PC (1999) Natural history of progression after PSA elevation following radical prostatectomy. JAMA 281(17):1591–1597
4. Bouchelouche K, Choyke PL (2018) Advances in prostate-specific membrane antigen PET of prostate cancer. Curr Opin Oncol 30(3):189–196

5. Yu T, Meng N, Chi D, Zhao Y, Wang Z, Luo Y (2015) Diagnostic value of dynamic contrast-enhanced magnetic resonance imaging in detecting residual or recurrent prostate cancer after radical prostatectomy: a pooled analysis of 12 individual studies. Cell Biochem Biophys 72(3):687–694
6. Wood DP Jr, Banks ER, Humphreys S, McRoberts JW, Rangnekar VM (1994) Identification of bone marrow micrometastases in patients with prostate cancer. Cancer 74(9):2533–2540
7. Schlimok G, Funke I, Holzmann B et al (1987) Micrometastatic cancer cells in bone marrow: in vitro detection with anti-cytokeratin and in vivo labeling with anti-17-1A monoclonal antibodies. Proc Natl Acad Sci U S A 84(23):8672–8676
8. Mansi JL, Berger U, Wilson P, Shearer R, Coombes RC (1988) Detection of tumor cells in bone marrow of patients with prostatic carcinoma by immunocytochemical techniques. J Urol 139(3):545–548
9. Melchior SW, Corey E, Ellis WJ et al (1997) Early tumor cell dissemination in patients with clinically localized carcinoma of the prostate. Clin Cancer Res 3(2):249–256
10. Wood DP Jr, Banerjee M (1997) Presence of circulating prostate cells in the bone marrow of patients undergoing radical prostatectomy is predictive of disease-free survival. J Clin Oncol 15(12):3451–3457
11. Braun S, Pantel K, Muller P et al (2000) Cytokeratin-positive cells in the bone marrow and survival of patients with stage I, II, or III breast cancer. N Engl J Med 342(8):525–533
12. Pantel K, Aignherr C, Kollermann J, Caprano J, Riethmuller G, Kollermann MW (1995) Immunocytochemical detection of isolated tumour cells in bone marrow of patients with untreated stage C prostatic cancer. Eur J Cancer 31A(10):1627–1632
13. Weckermann D, Muller P, Wawroschek F, Harzmann R, Riethmuller G, Schlimok G (2001) Disseminated cytokeratin positive tumor cells in the bone marrow of patients with prostate cancer: detection and prognostic value. J Urol 166(2):699–703
14. Kollermann J, Weikert S, Schostak M et al (2008) Prognostic significance of disseminated tumor cells in the bone marrow of prostate cancer patients treated with neoadjuvant hormone treatment. J Clin Oncol 26(30):4928–4933
15. Ellis WJ, Pfitzenmaier J, Colli J, Arfman E, Lange PH, Vessella RL (2003) Detection and isolation of prostate cancer cells from peripheral blood and bone marrow. Urology 61(2):277–281
16. Pfitzenmaier J, Ellis WJ, Hawley S et al (2007) The detection and isolation of viable prostate-specific antigen positive epithelial cells by enrichment: a comparison to standard prostate-specific antigen reverse transcriptase polymerase chain reaction and its clinical relevance in prostate cancer. Urol Oncol 25(3):214–220
17. Holcomb IN, Grove DI, Kinnunen M et al (2008) Genomic alterations indicate tumor origin and varied metastatic potential of disseminated cells from prostate cancer patients. Cancer Res 68(14):5599–5608
18. Morgan TM, Lange PH, Porter MP et al (2009) Disseminated tumor cells in prostate cancer patients after radical prostatectomy and without evidence of disease predicts biochemical recurrence. Clin Cancer Res 15(2):677–683
19. Chery L, Lam HM, Coleman I et al (2014) Characterization of single disseminated prostate cancer cells reveals tumor cell heterogeneity and identifies dormancy associated pathways. Oncotarget 5(20):9939–9951
20. Aguirre-Ghiso JA, Estrada Y, Liu D, Ossowski L (2003) ERK(MAPK) activity as a determinant of tumor growth and dormancy; regulation by p38(SAPK). Cancer Res 63(7):1684–1695
21. Murray NP, Reyes E, Tapia P, Badinez L, Orellana N (2012) Differential expression of matrix metalloproteinase-2 expression in disseminated tumor cells and micrometastasis in bone marrow of patients with nonmetastatic and metastatic prostate cancer: theoretical considerations and clinical implications-an immunocytochemical study. Bone Marrow Res 2012:259351
22. Murray NP, Aedo S, Fuentealba C, Reyes E, Salazar A (2018) Minimum residual disease in patients post radical prostatectomy for prostate cancer: theoretical considerations, clinical implications and treatment outcome. Asian Pac J Cancer Prev 19(1):229–236
23. Sosa MS, Bragado P, Aguirre-Ghiso JA (2014) Mechanisms of disseminated cancer cell dormancy: an awakening field. Nat Rev Cancer 14(9):611–622
24. Chalfin HJ, Glavaris SA, Malihi PD et al (2018) Prostate cancer disseminated tumor cells are rarely detected in the bone marrow of localized patients undergoing radical prostatectomy across multiple rare cell detection platforms. J Urol 199:1494–1501
25. Vandekerkhove G, Chi KN, Wyatt AW (2017) Clinical utility of emerging liquid biomarkers in advanced prostate cancer. Cancer Genet
26. Murray NP, Reyes E, Orellana N et al (2013) Secondary circulating prostate cells predict biochemical failure in prostate cancer patients after radical prostatectomy and without evidence of disease. ScientificWorldJournal 2013:762064
27. Kuske A, Gorges TM, Tennstedt P et al (2016) Improved detection of circulating tumor cells in non-metastatic high-risk prostate cancer patients. Sci Rep 6:39736
28. Markou A, Lazaridou M, Paraskevopoulos P et al (2018) Multiplex gene expression profiling of in vivo isolated circulating tumor cells in high-risk prostate cancer patients. Clin Chem 64(2):297–306

Minimal Residual Disease in Head and Neck Cancer and Esophageal Cancer

4

Christoph Sproll, Georg Fluegen, and Nikolas H. Stoecklein

Abstract
Malignant epithelial tumors of the upper digestive tract are a major cause of cancer-related death worldwide. The most common of these cancers are head and neck squamous cell carcinomas (HNSCC) and esophageal cancers (EC), which are both characterized by early dissemination and poor prognosis. Although patients with early detected cancers can be subjected to multimodal therapies with curative intention, they are endangered by lethal relapses that frequently occur. These relapses originate from minimal residual cancer (MRD) cells that can only be traced by highly sensitive molecular methods as rare disseminated tumor cells (DTC). The aim of this chapter is to comprehensively inform the reader about the detection, the mode of spread, the clinical relevance, and the biology of DTCs in HNSCC and EC. A better understanding of DTCs will be key to suppress progression of the upper digestive tract cancers more effectively.

Keywords
Head and neck squamous cell carcinoma · Esophageal cancers · TP53 · CCDN1 · MYC · EGFR · MRD · Bone marrow · Lymph node · DTC · Whole genome amplification · Whole trascriptome amplification

C. Sproll
Department of Oral- and Maxillofacial Surgery, Medical Faculty, University Hospital of the Heinrich-Heine-University Düsseldorf, Düsseldorf, Germany

G. Fluegen · N. H. Stoecklein (✉)
Department of General, Visceral and Pediatric Surgery, Medical Faculty, University Hospital of the Heinrich-Heine-University Düsseldorf, Düsseldorf, Germany
e-mail: Nikolas.Stoecklein@hhu.de

4.1 Introduction

Within the upper digestive tract occur some of the most aggressive cancer types characterized by early dissemination, metastasis and poor prognosis. The major group of squamous cell carcinomas arising from the mucosa of the oral cavity, oropharynx, hypopharynx, or larynx, which are subsumed as head and neck squamous cell carcinoma (HNSCC), account for 95% of cancers in the head and neck region. HNSCC is a leading cause of morbidity and mortality worldwide and is responsible for approximately 1.6% of all cancer-related deaths in the United States, with an estimated 49,670 new cases and 9700 deaths in 2017 [1, 2]. Another important group of upper digestive tract cancers can develop distal from the pharynx in the esophagus.

J. A. Aguirre-Ghiso (ed.), *Biological Mechanisms of Minimal Residual Disease and Systemic Cancer*, Advances in Experimental Medicine and Biology 1100,
https://doi.org/10.1007/978-3-319-97746-1_4

Esophageal cancers (EC) are the eighth most commonly diagnosed cancer entity and sixth leading cause of cancer-related deaths worldwide. In the year 2017, 16,940 estimated new cases of esophageal cancer were diagnosed in the US and 15,690 died from the disease [3] Like HNSCCs, ECs spread early to lymph nodes (LNs) but have a higher likelihood of more rapid systemic progression and development of distant metastases. Accordingly, the 5-year survival rates are quite low ranging from 15% to 25% [4]. There are two major histologic types of EC, esophageal squamous cell carcinoma (ESCC) and esophageal adenocarcinoma (EAC). The ESCC subtype largely shares the etiologic basis as well as many biological characteristics with HNSCC, e.g. similar molecular alteration patterns (including alterations in TP53, CCDN1, MYC, and EGFR) [5–7]. Moreover, synchronous and metachronous HNSCC/ESCC development can be observed due to the phenomenon of "field cancerization" caused by similar exposition to carcinogens from tobacco consumption and alcohol abuse [5, 8, 9]. The other major subtype, EAC, occurs mainly in the distal third of the esophagus and is markedly different from ESCC concerning etiology and biology. However, due to their similar metastatic behavior and their common intrinsic resistance to conventional systemic therapies we will consider them as one entity in this chapter [10].

Potentially curable HNSCC and EC are treated by surgery, radiation therapy, or both, which have to address the primary site and the lymphatic drainage to achieve locoregional control. However, despite modern surgical approaches with curative intent as well as advancements in the efficacy of (neo)adjuvant therapies, cancer cells can remain locally or at distant sites where they escape clinical diagnostics, survive and eventually proliferate. This so called minimal residual disease (MRD) can become the starting point of a local, regional or distant relapse.

The goal of this chapter is to inform the reader about the occurrence, detection, and clinical relevance of clinically occult disseminated cancer in HNSCC and EC, which is the presumed origin of metastatic relapse and is increasingly becoming a concern of clinical routine. We believe that future investigations should focus on the analysis of these early metastatic seeds to gain a better understanding of dissemination and metastasis, enabling more effective therapies to prevent metastatic outgrowth.

4.2 Clinical Relevance of MRD in HNSCC and EC

HNSCC and EC are both very aggressive cancers with a short-term lethal outcome in all untreated cases. Since HNSCC and EC often affect elderly patients with significant comorbidities [11] and present frequently at advanced tumor stages [12], the lifespan after diagnosis is usually quite limited [13]. Therefore, long clinical dormancy periods, as e.g. frequently observed in breast or prostate cancer, are less common. However, during the last two decades patients increasingly get stratified to more effective multimodal therapies with curative intention, resulting in prolonged relapse free intervals and higher incidences of metastatic relapses. This may also be in part due to the rise in EAC in Northern Europe and in North America in the last two decades, which is more amenable to early detection and displays longer survival rates after surgical multimodal therapy, while ESCC remains the most common histological subtype worldwide [14, 15]. In case of HNSCC, there has been an increase in incidence of oropharyngeal carcinoma associated with HPV–infections during the last decade that occurs mainly in younger Caucasian males presenting with small primary tumors, advanced nodal disease [16] and overall better prognosis [17]. In general, patients with early stage HNSCC (UICC I–II) devoid of clinically apparent LN or distant metastases may be cured with modern surgical approaches, with 5-year survival rates of over 80% [18]. Nevertheless, about 20–30% of these patients with supposed locally confined tumors develop locoregional relapse and/or distant metastases [19]. In analogy to early stage HNSCC, the reported 5-year survival rates of over 70% for UICC I/II EC as well as cancers that responded completely to neoadjuvant therapy (ypT0pN0cM0-EC) are quite high, but patients are threatened by life-limiting metastatic relapses at distant sites [20]. From a clinical perspective, the occurrence of distant metastases in HNSCC or EC after initially successful treatments changes care from curative to palliative regimes. In contrast to multimodal curative settings, systemic therapies are quite ineffective in palliative situations, due to the intrinsic therapeutic resistance of heterogeneous HNSCC/EC cells and simply because the lesions are often too large for effective medical treatment (10^{10}–10^{11} cells) [21]. This clearly highlights the importance to detect and investigate MRD cells in HNSCC and EC to understand their biology and to enable more effective strategies to prevent their outgrowth.

4.3 Mode of Spread in HNSCC and EC

In order to understand where disseminated MRD can persist and can develop into metastasis we will briefly review the metastatic pattern of HNSCC and EC. Importantly, both tumor entities have a strong tendency to initially target regional LNs, possibly due to the high density of lymphatic vessels in the oral cavity, the pharynx and the esophagus. LN metastases are relevant because they are the single most relevant predictor of outcome for both cancer types. In case of HNSCC, regional LN involvement reduces the 5-year survival rate by nearly 50% [22]. Proper staging and treatment of the cervical LNs became therefore a cornerstone of HNSCC management. Hence, precise LN mapping is required and first attempts to classify the complex anatomy of cervical LNs were made by Rouviére and relied on anatomical landmarks [23–25]. This anatomical system has been challenged over time by a simplified pathophysiological classification for clinical use and originated from the work of Lindberg, who analyzed patients with HNSCC between 1948 and 1965 at M.D. Anderson in order to determine the topographical distribution of suspected LN metastases. He defined nine nodal regions on each side of the neck and of the parotideal LNs. On the basis of histopathological data from ND specimens [26, 27] this scheme was further developed to the current mapping system, which was last modified in 2002, comprising now six regions per neck side [28]. This system was commissioned by the AAHNS (American Academy of Otolaryngology – Head and Neck Surgery) and is recommended by the AJCC [29]. The easy accessibility and the frequent detection of occult metastatic disease before the onset of the era of modern imaging techniques prompted a very radical removal of the LNs, called radical neck dissection (ND), at the beginning of the last century, which became the gold standard in neck treatment for many years [30]. However, an evolution of ND towards less invasive procedures took place: modified radical ND (type I–III) [31], selective ND [28], super-selective ND and sentinel LN biopsy (SLNB) were developed. The recognition that the prognosis is not commensurate with the extent of ND [32] and the steady improvement of imaging modalities suggested diagnostic ND in clinically negative necks (cN0) obsolete. But the results of a recently published prospective clinical trial highlighted the importance of elective ND not only for eliminating occult MRD but also to guide adjuvant radiotherapy [33].

Similar to HNSCC, locoregional LN involvement is of strong prognostic significance for survival and relapse rates in EC [34]. It is important to note the very complex lymphatic anatomy of the esophagus with two embryologically distinct lymphatic drainage pathways, one located in the submucosa and the other between the two muscle layers of the muscularis propria [35]. The submucosa drains axially into the paracardial and laryngeal LNs, then into the celiac and supraclavicular LNs [36]. The intermuscular drainage path of the mid and lower esophagus drains horizontally into the paraesophageal LNs of the middle and lower mediastinum first [35].

From a surgical point of view and with regard to the extent of LN dissection, the regional LNs of the esophagus are divided into three fields: cervical, mediastinal and perigastric. Within these fields, a subdivision is made between several LN stations that have been adopted from the Japanese Classification of Esophageal Cancer [37]. Altogether, 26 different LN positions are defined by anatomical landmarks, by the field, and a number (e.g. thoracic LN, 106recR: right recurrent nerve LN) [38]. In EC, the lymphatic drainage and thus the likelihood of LN metastasis of the tumor depends on the axial position and longitudinal extent of the tumor, on the depth of invasion (pT) as well as on the differentiation of the tumor [39]. In its axial extension the esophagus is divided into three parts, but cross-border tumors and the crucial proximal boundary of the tumors further complicate the prognostication of LN metastasis [40]. However, due to the higher anatomical complexity strict mapping like in HNSCC with its easy accessible LNs is hardly achievable and is not routinely performed in EC. For effective staging, significantly more LNs need to be therefore dissected in EC compared to HNSCC in order to reliably determine a pN0-status [41]. Also, due to the unpredictable lymphatic drainage, a sentinel LN biopsy (SLNB) is not practical in EC. For example, a SLNB study in EC revealed an average of 4.7 SLNs detected all over from the neck to the abdominal area independent from the localization of the primary tumor [42]. The required number of resected LN to remove all relevant seeded LNs and to maximize survival depends on the invasion depth of the cancer (pT-category): 10 LNs are required in pT1-tumors, 20 in pT2-tumors and at least 30 in pT3- or pT4-tumors [29]. This explains that unlike in the HNSCC, LN-dissection has progressively evolved towards increasing radicalness, since locoregional recurrences were more common in limited LN dissections [38]. In this context, several authors suggest the so-called radical 3-field lymphadenectomy (3FL) of the bilateral cervical, mediastinal, and abdominal regions. This takes into account the independence of the anatomical distance from the primary tumor and the rather complex lymphatic drainage system [38]. However, due to higher complication rates and heterogeneous outcomes between studies, it is still under debate whether 3FL is required or a 2-field lymphadenectomy is sufficient [38, 43]. However, marked differences between diagnostic and therapeutic LN-dissection like in HNSCC have finally not been reported in EC.

While LN metastases are the most important prognostic factor in potentially curable patients, life-expectancy becomes dramatically shortened when distant metastatic relapses arise. In order to understand, at which distant sites MRD cells can find the most fertile soil, we briefly review data revealing the metastatic pattern in HNSCC and EC. Such data came initially from post mortem studies, because clinical (functional) cross-sectional imaging was lacking. In HNSCC, the largest such series was published 1923 by Hitchings as a literature review for George Crile, reporting a prevalence of metastases in 4.500 HNSCC patients of less than 1% [30]. This report was the basis for the influential concept that HNSCC is almost exclusively a locoregional disease and the impenetrability of the collar of lymphatics around the neck was proposed as metastatic barrier [30]. However, autopsy studies from the 1970s to 1980s including patients treated with curative intent [44–46] revealed distant metastases in 12–25%, especially in patients with higher T-stages. These data also indicated (for the first time) that dissemination of metastatic cells must have occurred before the initial treatment [45]. Although meticulous investigation revealed that almost no organ was spared from metastasis, the predilection sites were usually lungs, bones, liver, skin, and brain [45]. Within a median 2-year interval around 15% of the M0-patients who received treatment with curative intention in M0-patients will develop distant disease with or without locoregional recurrence [46, 47]. In case of EC, distant metastasis is more common and associated with a worse prognosis than in HNSCC. Only around 5% of patients with distant metastases are still alive after 5 years in EC compared to 38.5% in HNSCC [3]. Similar to HNSCC, the main metastatic sites in both subtypes of EC are the liver, lungs, bones, and brain, yet there is a difference in the frequency of metastasis of these sites. While EACs metastasize more frequently to the liver (19.3 vs. 8.9%), bone (8.6 vs. 5.6%) and brain (2.2 vs. 0.6%), ESCC target more frequently the lung (11.4 vs. 9.1%) [4].

4.4 Methods to Detect MRD in HNSCC and EC

In this chapter, we disregard cancer cells potentially remaining at the resection margin and refer to MRD only as the clinically occult tumor cell population that has spread from the primary tumor site into regional LNs as well as to distant sites. The detection of minimal residual cancer cells in patients with operable epithelial malignancies is technically quite challenging (i) because of their extremely low frequency and (ii) due to the lack of cancer-specific markers for detection. To overcome the latter problem, sensitive assays have been developed targeting the (remnant) epithelial phenotype of the cancer cells within mesenchymal organs [48]. To screen for the presence of epithelial MRD cells, two mesenchymal indicator organs have been established: (regional) LNs and BM. LNs are the most frequent target for clinically overt metastasis in HNSCC/EC. In the BM, on the other hand, clinically detectable metastases are relatively less frequent in upper digestive tract cancers. BM is thus merely considered as a probing site for hematogenously disseminated epithelial tumor cells as indicators for systemic disease [49, 50]. In contrast to peripheral blood, BM provides a supportive environment for disseminated cancer cells, sustaining long time survival and enabling tumor cell dormancy for years to decades [51]. The detected MRD cells are currently usually designated as disseminated tumor cells (DTCs), indicating that these viable cells have arrived at a site distant form the primary tumor, but have not formed a metastasis (yet). In the current literature, the nomenclature of the clinical occult disseminated disease is not standardized and some authors refer to these cells also as disseminated cancer cells (DCCs), isolated tumor cells (ITCs), or micrometastatic disease. However, a clear guideline for morphological evaluation, especially for LNs, can be found in the UICC staging manual. Here, single cells or cluster of cells with a size less than 0.2 mm are defined as ITCs. According to this definition, ITCs are usually characterized by the absence of evidence for metastatic activity, such as proliferation and/or stromal reaction. Malignant clusters and metastatic lesions with a size between 0.2 and 2.0 mm are denoted as micrometastasis (MM). This classification has been first published by Hermanek et al. [52] and was later integrated into the 6th edition of the TNM classification for malignant tumors (UICC) [50].

4.5 Immuno-Detection of LN-DTCs

Immunocytochemistry (ICC) using anti-epithelial antibodies is the most commonly applied method to detect LN-DTCs. While earlier studies from the 1990s more often used snap frozen material of selectively sampled lymph nodes, the majority of studies applied IHC on slides of FFPE (formalin fixed and paraffin embedded) samples that had been already evaluated by routine histopathology (see Table). The advantage of this approach is that samples from standardized lymphadenectomies can be reevaluated, providing a more representative overview of the minimal tumor load overlooked by routine pathology. In addition, it elicits an impression of the minimal residual disease that would have been left behind in non-resected LNs. Furthermore, certain morphological changes indicative of micrometastatic disease can be detected in experienced hands like the formation of desmoplastic stroma [53] or focal capsular thickening [54]. Most groups have used standard avidin-biotin-complex IHCs for FFPE tissues combined with pancytokeratin antibodies (Table). The number of the analyzed slides varies, but often only one additional slide was analyzed. Only in few studies serial step sections were processed similar to the analysis of SLN biopsies [55–58]. In order to obtain comprehensive genomic information on LN-DTCs, Klein and colleagues described a method to generate single cell suspensions from a part of the fresh LN by mechanical disaggregation. The cells are then plated on adhesive slides at a density of 250,000 cells per 227 mm^2. After immuno-staining, intact single DTCs can be isolated from the slide by micromanipulation and subjected to whole genome amplification (WGA) and further genomic analysis [59, 60].

The immuno-assays to detect lymphatic DTCs are very sensitive and, depending on the antibodies, highly specific. However, the staining protocols are not standardized, which might contribute to some heterogeneity between studies. Cytokeratins (CKs), components of the cytoskeleton of all epithelial cells [61] were predominantly used as epithelial antigen (Tables 4.1 and 4.2). Although LNs provide an exclusively mesenchymal background and resident cells do not normally express cytokeratins, interstitial reticulum cells can occasionally express CK8 and CK18. This expression was observed in reactive, inflammatory and neoplastic lymphadenopathies. Due to their different affinities and specificities, the diverse anti-cytokeratin antibodies can react differently with this low CK expression of the reticulum cells. It was repeatedly demonstrated that compared to the antibody clone CAM5.2, the more frequently used clone AE1/AE3 reacts only weakly with this cell type although it also recognizes cytokeratin 18 [62, 63]. However, since reticulum cells display a typical dendritic morphology they are discernable from epithelioid cancer cells. Plasma cells were also reported to react with anti-cytokeratin antibodies, but these cells also have a characteristic morphology, enabling exclusion of these [63, 64]. Salivary gland inclusions are a common phenomenon especially in neck LNs. They often present as acinar cells with CK-positive mucus content and in the form of annular shaped structures and can morphologically easily be discerned from metastatic cells [65]. To circumvent unspecific staining resulting e.g. from reticulum cells, some investigators used the Ber-EP4 antibody directed against EpCAM (CD236) [66], which was reported to have a higher specificity in lymphatic tissue for the detection of epithelial tumor cells [34, 35]. However, the potential disadvantage of EpCAM, a 40-kDa transmembrane glycoprotein expressed on the surface of most epithelial tissues, is its inconsistent expression on some cancer types, including HNSCC and EC (Tables 4.3 and 4.4).

4.6 Immuno-Detection of BM-DTCs

For the detection of hematogenously spread DTCs, usually 10–20 milliliters of BM are aspirated from the pelvic bones. To avoid cross-contamination with epidermal cells, a skin incision before the biopsy puncture is obligatory. A few studies in EC patients have also investigated rib BM [67, 68]. This is easily accessible during thoracotomy for esophageal resection, when a rib segment is excised to facilitate rib-retraction. From this resected bone, the BM is flushed out with cell culture medium and then further processed. In order to efficiently detect DTCs in BM samples, an enrichment step is required to reduce the background of hematopoietic cells. Up to now, the most widely applied enrichment method for BM samples is a density gradient centrifugation using an aqueous solution of modified highly branched hydrophilic polysaccharides (Ficoll-Hypaque). Upon centrifugation, the different BM cell types migrate according to their buoyant density into distinct layers, with the mononuclear cell (MNC) fraction containing the DTCs localized at the interphase between plasma and the polysaccharide solution [69]. For subsequent DTC detection, cytospins are generated from the enriched cell suspensions and subjected to immunocytochemistry (ICC) using anti-epithelial antibodies. The most commonly used monoclonal antibody in the field has been the clone A45/BB3 (Micromet, Munich, Germany), reacting with common epitopes of several cytokeratins including CK8, CK18 and CK19. This antibody has been also used to study DTCs in upper digestive tract cancers along the clone AE1/AE3 that covers a large spectrum of cytokeratins (CK10, CK14–16, CK19 and CK1–8). For visualization, an alkaline phosphatase–anti-alkaline phosphatase (APAAP) ICC with levamisole as blocking reagent, new fuchsin as chromogen and hematoxylin for nuclear counter-staining is a highly recommended system and has been tested in inter-European ring trials [70]. However, due to the fixation, this assay is not compatible with downstream molecular analysis of DTCs. Therefore, protocols for fresh cell

Table 4.1 Studies on LN-DTCs in EC

Author	Year	Ref.	Patients (all)	UICC/ AJCC-stages	No LN (tumor free)	Material	Antibodies	Targets	Definition of MM and DTCs	Pat (all) pos	Pat (all) pos %	No LN pos	LN pos. %	Prognostic	Type	pN0 Pat	pN0 Pat pos	pN0 pat pos%
Izbicki	1997	[99]	68	I,II,III	399	Fresh Frozen	Ber-EP4		Isolated Ber-EP4–positive cells or cell clusters (containing up to three cells) in sinuses, lymphoid interstitium, or both	42	62	67	16,8	Yes	mult	30	15	50
Natsugoe	1998	[100]	69	0,I,II,III	1954	FFPE	AE1/AE3	MM + TCM	1) TCM, tumor cells or clusters with- out a surrounding stromal reaction; 2) MM, tumor cells or clusters with a stromal reaction smaller than 0.5 mm in greatest dimension	29	42	46	2,4	Yes	univ	41	13	31,7
Glickman	1999	[101]	78	0,I,II,III	574	FFPE	AE1/AE3		one or more strongly immunoreactive epithelial cells in the subcapsular sinus or in the cortex of the LN < 2 mm	20	26	40	7,0	NS	mult	41	11	nr

(continued)

Table 4.1 (continued)

Author	Year	Ref.	Patients (all)	UICC/ AJCC-stages	No LN (tumor free)	Material	Antibodies	Targets	Definition of MM and DTCs	Pat (all) pos	Pat (all) pos %	No LN pos	LN pos. %	Prognostic	Type	pN0 Pat	pN0 Pat pos	pN0 pat pos%
Mueller	2000	[142]	145	I,II,III	3987	FFPE	AE1/AE3	MM + TCM pN0 11/75	1) TCM, defined as free-floating tumor cells or cell clusters within the subcapsular sinus or intramedullary sinuses of the LN; 2) MM, defined as firmly anchored or attached tumor cells or clusters 0.5 mm in greatest dimension within the LN;	48	33	196	4,9	YES (MM)	univ	75	11	14,7
Matsumoto	2000	[102]	59	I,II	2714	FFPE	AE1/AE3		MM: clear cytoplasmic immunoreactivity of single cells or several scattering single cells	39	66	66	2,4	Yes	univ	59	39	66,1
Komukai	2000	[103]	37	I,II	2774	FFPE	AE1/AE3		MM: either a single cell or a cluster of 5 or fewer CK-positive cells	14	38	29	1,0	Yes	mult	37	14	37,8
Sato	2001	[104]	50	I,II	1840	FFPE	AE1/AE3		AE1/AE3 positive single cells, multiple individual cells, clusters or multiple clusters not found by H&E	20	40	29	1,6	NS	mult	50	20	40
Qubain	2001	[105]	86	I,II,III,IV	2203	FFPE	AE1/AE3	MM	MM: Tumor involvement of the LNs detected using IHC alone	16	19	23	1,0	NA	na	53	9	nr

Komukai	2002	[106]	104	I,II,III	7769	FFPE	AE1/AE3		MM: either a single cell or a cluster of 5 or fewer CK-positive cells	47	45	106	1,4	Yes	mult	38	14	nr
Hosch	2001	[88]	86	I,II,III	540	Fresh	Ber-EP4		Minimal tumor cell involvement in a LN that was considered tumor-free by routine histopathologic staining was defined as the presence of up to three Ber-EP4–positive cells within the body of the LN	61	71	97	18,0	Yes	mult	42	26	62
Doki	2002	[107]	41	II	2148	FFPE	AE1/AE3	MM + CD (w/o nuclei)	MM was defined as a single isolated or cluster of cancer cells, stained by CK- antibody	18	44	42	2,0	Yes (MM)	univ	41	18	43,9
Vazquez-Sequeiros	2002	[108]	124	0,I,II	2001	FFPE	AE1/AE3		true positive, AE1/AE3 staining of cells morphologically compatible with carcinoma	14	11	na	na	No	univ	124	14	11,3
Waterman	2004	[109]	28	I,II,III,IV	1970	FFPE	AE1/AE3	pN0 & <10 LNM ratio	individual immuno-reactive cells (ie, expressed CK-antigens) or clusters of cells were present within the substance of the LN and only if the morphology of the cells was consistent with cancer	17	61	na	na	No	univ	8	3	37,5

(continued)

Table 4.1 (continued)

Author	Year	Ref.	Patients (all)	UICC/ AJCC-stages	No LN (tumor free)	Material	Antibodies	Targets	Definition of MM and DTCs	Pat (all) pos	Pat (all) pos %	No LN pos	LN pos. %	Prognostic	Type	pN0 Pat	pN0 Pat pos	pN0 pat pos%
Heeren	2005	[110]	60	I,II	524	FFPE	AE1/AE3	ITC, clusters	The presence of either isolated (less than three staining cells) or cluster type, i.e. clusters of three or more stained cells with IHC techniques on previously considered N0 LNs, was considered as nodal MM.	18	30	126	24,0	Yes	univ	60	18	30
McGuill	2008	[111]	146	0,I,II, III	1694	FFPE	MNF116	ITC	ITC: if they contained single or small clusters of strongly immunoreactive epithelial cells in the subcapsular sinus or in the cortex of the LN	12	8	24	1,4	Yes	mult	146	12	8,2
Buskens	2008	[112]	20	I,II	578	FFPE	Ber-EP4, AE1/AE3, CAM5.2	ITC/MM	ITCs and small clusters of positive cells (< 0.2 mm) were discriminated from MMs (0.2–2 mm). Clusters of positive cells with malignant characteristics were only designated as MMs when detected in the sinuses or lymphoid interstitium.	5	25	16	2,8	Yes	univ	20	5	25

Zingg	2009	[113]	86	0,I,II	1204	FFPE	Lu-5	ITC / MM / ACPM	Hermanek classification	14 (ITC + MM)	16	50	4,2	Yes (SCC)	univ	86	14	16,3
Koenig	2009	[114]	73	I,II,III	1958	FFPE	AE1/AE3		Minimal tumor cell involvement in a LN that was considered to be tumor free by conventional histological staining was defined as the presence of one to ten positive cells in the body of the LN	25	34	47		Yes (ratio)	mult	33	9	27,2
Chao	2009	[115]	52	only pCR (II,III before CRT	510	FFPE	AE1/AE3		MM: AE1/AE3 positivity in two sections of a LN after evaluation by two independent pathologists	6	12	na	na	NS	univ			
Thompson	*2010*	*[116]*	*119*	*I,II,III*	*NA*	*FFPE*		*ITC/MM*	*Hermanek – classification*	*30*	*25*	*na*	*na*	*Yes*	*mult*	*119*	*30*	*26,9*
Grotenhuis	2010	[75]	63	I	NA	FFPE	CAM5.2	m2/m3/ sm1-tumors	Hermanek – classification	12	19	na	na	na	na	63	12	19
Prenzel	2012	[76]	48	I	1344	FFPE	AE1/AE3	pT1	Positive staining for AE1/AE3 of LNs was considered as micrometastatic involvement of these LN if at least three sections of one LN were positive after evaluation by two independent investigators	7	15	18	1,3	Yes	univ	48	7	14,6

(continued)

Table 4.1 (continued)

Author	Year	Ref.	Patients (all)	UICC/ AJCC-stages	No LN (tumor free)	Material	Antibodies	Targets	Definition of MM and DTCs	Pat (all) pos	Pat (all) pos %	No LN pos	LN pos. %	Prognostic	Type	pN0 Pat	pN0 Pat pos	pN0 pat pos%
Driemel	2013	[85]	28	I,II,III,IV	na	Fresh	Ber-EP4		clear K-positive cells	14	50	na	na	na	na			
Yanagi	2018	[117]	41	0,I,II,III,IV	1741	FFPE	AE1/AE3	nRCT vs no nRCT	1) TCM, tumor cells or clusters with- out a surrounding stromal reaction (Fig. 1); 2) MM, tumor cells or clusters with a stromal reaction smaller than 0.5 mm in greatest dimension	22	54	52	3,0	Yes (with pN)	univ	14	5	35,7

Table 4.2 Studies on LN-DTCs in HNSCC

Author	Year	Ref.	No of pts.	pN-status	UICC/AJCC-stages	No of LN	LN per pt.	Total no. of sections	Material	Processing of LN	Method	Antigen (method)	Def. of MM and DTCs	MMs			DTCs		Median follow up (months)	Prognostic	Type	Remarks
														Pos. LN (%)	Pos. levels AAHNS	Pos. pts. (%)	LN pos. for DTCs	Pos. Pts. for DTCs				
Ambrosch	1995	[55]	60	pN0	I, II, III, IV	1.020	17	5.999 H&E, 1.261 IHC	FFPE	step serial sectioning at 1 µm: every 10th slide: H&E, every 50th slide: IHC	H&E, IHC	pan-CK (APAAP)	≤ 2 mm diameter	8 (0,8%)	II	6 (10%)	NA	NA	37	NR	univ	Preoperative staging by palpation and ultrasound, MR, CT and FNAB not routinely applied, Micrometastasis ≤2 mm
Enepekides	1999	[118]	44	pN0	I, II, (IV)	419	12,3	NR	FFPE	NR	H&E, IHC	AE1/AE3 (Ventana™, immuno-peroxidase automated procedure)	Occult metastasis: missed on H&E-staining, detected by more sensitive method	4 (0,8%)	NR	4 (10%)	NA	NA	40,5 ± 3,5	DFS, OS: n.s.	univ	1/4: MET from papillary thyroid cancer; micrometastasis not defined
Yu	1999	[119]	46	pN0	NA	NR	NR	NR	FFPE	NR	H&E, IHC	pan-CK (AE1/AE3), NR	NR	13 (NR)	NR	11 (23,9%)	NA	NA	NR	NR	NA	Article in Chinese: only abstract available
Kocatürk	2003	[120]	22	pN0	I, II, III, IVA	512	23,3	NR	FFPE	cross-sections at 1, 2 or 3 levels depending on LN size, IHC at deepest cross section	H&E, IHC	pan-CK, NR	Occult metastasis: missed on H&E-staining, detected by more sensitive method	NR	NR	3 (13,63%)	NA	NA	mean 13 (3–36)	local recurrence, locoregional recurrence: n.s.	univ	Only supraglottic larynx included; also cN1 and cN2-pts. Included
Garrel	2006	[57]	18	cN0	I, II	66 SLN	3,7	NR since LN size unknown (40 sections per 1 cm LN)	Snap frozen	step serial sectioning: 2 × 5 µm every 250 µm	H&E, IHC	CK5,14,17 (Ventana™ DAB detection kit)	Hermanek-classification	9 (13,6%)	IA, IB, IIA, III, IV	7 (41,18%)	12 SLN (18,2%)	5 pts. pN0(i+)(sn) (29,41% of all pts., 45,4% of pN0-pts.)	NR	NR	NA	IHC used to validate RT-PCR for CK5, 14 and 17 of SLN; also pN + −patients (11 pN0, 4 pN1, 2 pN2b, 1 pN2c); micrometastasis ≤2 mm

Table 4.2 (continued)

Author	Year	Ref.	No of pts.	pN-status	UICC/AJCC-stages	No of LN	LN per pt.	Total no. of sections	Material	Processing of LN	Method	Antigen (method)	Def. of MM and DTCs	MMs			DTCs		Median follow up (months)	Prognostic	Type	Remarks
														Pos. LN (%)	Pos. levels AAHNS	Pos. pts. (%)	LN pos. for DTCs	Pos. Pts. for DTCs				
Solassol	2010	[58]	22	cN0, 16 pN0, 4 pN1, 2 pN2b	I, II, III, IV	78 SLN	3,5	NR since LN size unknown (40 sections per 1 cm LN)	Snap frozen	step serial sectioning: 2 × 5 μm every 250 μm	H&E, IHC	pan-CK (AE1/AE3) (Ventana™ DAB detection kit)	Hermanek-classification	9 (11,54%)	IB, IIA, IIB, III, IV	6 (27,27%)	12 SLN (15,4%)	6 pts. pN0(i+)(sn) (27,3% of all pts., 37,5% of pN0-pts.)	NR	NR	NA	IHC used to validate RT-PCR for PVA[§], SSCA* and CK17 of SLN; also pN + −patients (16 pN0, 4 pN1, 2 pN2b)
Broglie	2013	[56]	111	cN0, 69 pN0, 8 pN0(i+)(sn), 15 pN1(mi)(sn), 2 pN2b(mi)(sn), 5 pN1(sn), 11 pN2b(sn), 1 pN2c(sn)	I, II, III, IV	392 SLN	3,6	NR since LN size unknown	FFPE	nodes cut into slices of 2 mm if >5 mm, embedded as a whole if ≤5 mm in max. diameter, then further step-sectioning in a superficial, middle and deep layer	H&E, IHC	pan-CK, NR	Hermanek-classification	NR	I-IV ipsilaterally, I-III contra-laterally	19 (17,11%)	NR	10 (9%)	37 (2–83)	OS, DSS, DFS significantly shorter for MMs, n.s. for ITCs	univ and mult	survival assessed at 3 years; 5-year rates NR
Sproll	2017	[87]	50	cN0 + pN0	I, II, III, IVA	1.137	22,7	4.328	FFPE	nodes cut into halves if >5 mm, embedded as a whole if ≤5 mm in max. diameter, then 4 consecutive slides of 4 μm	H&E, IHC	CKs 5/14, pan-CK (AE1/AE3), CD44v6; Vectastain ABC-Kit	Hermanek-classification	6 (0,52%)	IB, IIA, III	5 (10%)	41 (3,6%)	12 (24%)	71	DFS, OS: n.s.	univ	micrometastasis ≤2 mm, DTCs

Table 4.3 Studies on BM-DTCs in EC

Author	Year	Ref.	Patients (all)	UICC/ AJCC	Donor site	Antibody	Number of cells screened	Pat (all) pos	Pat (all) pos %	pN0 Pat	pN0 Pat. pos.	Prognostic	Type	Remarks
Thorban	1996	[121]	90	I,II,III,IV	Iliac Crest Aspirate	KL-1	4 × 10^5	37	41	39	15	Yes	univ	
O'Sullivan	1999	[67]	50	I,II,III	Rib flush	anti-CK18	0,5 × 10^6	44	88	18	16	NA		
O'Sullivan	1999	[67]	27	I,II,III	Iliac Crest Aspirate	ant-CK18	0,5 × 10^6	4	15	NR	NR	NA		
Thorban	2000	[122]	75	I,II,III,IV	Iliac Crest Aspirate	A45-B/B3	2 × 10^6	29	39	NR	NR	Yes	mult	neoadjuvant stage III
Bonavina	2001	[123]	48	I,II,III	Rib flush	AE1/AE3	7 × 10^5	38	79	NR	NR	NA		
Bonavina	2001	[123]	18	I,II,III	Iliac Crest Aspirate	AE1/AE3	7 × 10^5	2	11	NR	NR	NA		
Spence	2004	[124]	49	I,II,III,IV	Iliac Crest Aspirate	CAM5.2, AE1/AE3	4 × 10^6	22	45	42	19	NA		
Ryan	2004	[125]	106	I,II,III,IV	Rib flush	anti-CK18	10^6	57	54	71	31	NA		CRT:42%, –CRT:67%
Kaifi	2005	[126]	136	I,II,III,IV	Iliac Crest Aspirate	A45-B/B3	NR	60	44	63	NR	Yes	mult	
Stoecklein	2008	[60]	107	I,II,III,IV	Iliac Crest Aspirate	A45-B/B3	10^6	44	41	52		n.s.	univ	
Bagheri	2011	[127]	43	I,II,III	rib / slide	MNF	NA (10 microscopic fields 10 × 40)	13	30			NA		
Gray	2012	[128]	42		Iliac Crest Aspirate	CAM5.2, AE1/AE3	4 × 10^6	19	45			No		
Vashist	2012	[129]	370	I,II,III,IV	Iliac Crest Aspirate	A45-B/B3	2 × 10^6	120	32			Yes	mult	
Driemel	2014	[85]	83	I,II,III,IV	Iliac Crest Aspirate	A45-B/B3	10^6	21	25			No		
Ryan	2015	[130]	88	0,I,II,III	Rib flush	anti-CK18	10^6	47	53	29		Yes	mult	

Table 4.4 Studies on BM-DTCs in HNSCC

Author	Year	Ref.	Pat (all)	pN-Status	UICC-stages	Donor site	Method	Antibody	Number of cells screened	Pat (all) pos	Pat (all) pos %	pN0 pat.	pN0 pat. pos	pN0 pat. Pos %	pN+ Pat	pN+ Pat pos	pN+ Pat pos %	Prognostic	Type	Pos. correlation	No correlation	Remarks
Wollenberg	1994	[131]	108	pN0 (49), pN+ (58)	I (19), II (14), III (31), IV (44)	Iliac crest aspirate	ICC	anti-CK19	2×10^6	41	37,96	49	18	36,7	58	22	37,9	Yes	mult	UICC-stage, recurrence in stages III and IV	T-, N-, G-Status	
Gath	1995	[132]	31	pN0 (17), pN1 (4), pN2 (8), pN3 (1)	I (6),II (1), III (5), IVA (18), IVB (1)	Iliac crest aspirate	ICC	A45-B/B3	1×10^6	10	32,26	17	5	29,4	13	5	38,5	NA	mult	G-status	pT-Status, pN-Status	
Gath	1999	[133]	149	NR	NR	Iliac crest aspirate	ICC	A45-B/B3	1×10^6	30	20,13	NR	NR	NR	NR	NR	NR	NA	mult	NA	NA	Announced paper in Gath et al. 1999, not yet published
Partridge	1999	[134]	17	NR	NA	Iliac crest aspirate	ICC (positive IMS)	BerEp4 + A45/B/B3 (positive selection)	2×10^7	2	11,76	NR	NR	NR	NR	NR	NR	NA	NA	NR	NR	Methodological paper effects of positive and negative selection. No clinical follow up data provided.
		[134]	17	NR	NA	Iliac crest aspirate	ICC (negative IMS)	anti-CD45 + AE1/AE3 (negative selection)	2×10^7	11	64,71	NR	NR	NR	NR	NR	NR	NA	NA	NR	NR	
Chaubal	1999	[135]	15	NR	I (2), III (2), IV (11)	Fresh frozen and Iliac crest aspirate	ICC	C215		7	46,67	NA	NA	NA				NA	NA	NA	NA	
Partridge	2000	[136]	18	pN0 (8), pN1 (5), pN2 (4), pN3 (1)	II (5), III (4), IVA (8), IVB (1)	Iliac crest aspirate	ICC	anti-CD45 + AE1/AE3	2×10^7	6	33,33	8	0	0	10	6	60	NA	NA	NA	NA	

Partridge	2003	[137]	40	pN0 (9), pN1 (13), pN2 (13), NA (3)	0 (1), I (4), II (4), III (9), IVA (18), NA (4)	Iliac crest aspirate pre-OP	ICC (negative IMS)	anti-CD45 + AE1/AE3	2 × 10^7	14 (2 isotype controls positive)	35	9	(1, isotype control positive)	11,1	26	11 (1 isotype control positive)	42,31	Yes	univ	DMFS, DFS		
		[137]	31	pN0 (9), pN1 (13), pN2 (13), NA (3)	0 (1), I (4), II (4), III (9), IVA (18), NA (4)	Iliac crest aspirate 3–4 month post-OP	ICC (negative IMS)	anti-CD45 + AE1/AE3	2 × 10^7	2	6,45	9	0	0	26	2	7,69	No	univ	DMFS, DFS		
Wollenberg	2004	[138]	176	pN0 (88), pN1 (31), pN2 (50), pN3 (7)	I and II (60), III and IV (116)	Iliac crest aspirate	ICC	anti-CK19	2 × 10^6	54	30,7	88	27	30,7	88	27	30,7	Yes	mult	relapse (locoregional +/− distant), OS	UICC-stage, pT-stage, pN-stage, localisation, grading	misleading title. But paper is about DTCs; article in german, follow-up 60 months;
Bleijerveld	2011	[139]	20	NR	II (4), III (1), IV (15)	Iliac crest aspirate	ICC, NR	NR	NR	10	50	NR	NR	NR	NR	NR	NR	NA	NA	NA	NA	BM-pos. vs. BM-neg. Patients only used to stratify patients for potentially DM-pos. vs. DM-neg. pts. Goal of the study to evaluate protein-signatures associated with tumor cell dissemination
Graveland	2012	[140]	86 (76)	pN0 (55), pN+ (46)	I and II (27), III and IV (74)	Iliac crest aspirate	ICC	A45-B/B3	NR	11	14,5	NR	NR	NR	NR	NR	NR	No	univ	none	locoregional RFS, DFS, OS, DSS	Main goal of the study was to evaluate MRD in deep surgical margins
Gröbe	2014	[141]	90	pN0 (44), pN1 (22), pN2 (21), N3 (3)	I (22), II (10), III (27), IV (31)	Iliac crest aspirate	ICC	A45-B/B3	2 × 10^6	18	20	44	7	15,9	46	9	19,6	Yes	mult	pN-, M-status	pT-stage, UICC-stage	

preparations have been established [60, 71]. Also, with this assay false positive findings can occur and isotype controls are mandatory: in a meticulous study Borgen et al. uncovered direct reaction of endogenous alkaline phosphatase in plasma cells as another reason for unspecific positive cell staining in BM from healthy individuals [72].

4.7 Molecular Detection of DTCs

Another approach to detect occult DTCs is their molecular detection by quantitative real-time reverse-transcription PCR (qRT-PCR). The sensitivity for detection of cancer cells in a background of non-cancer cells can be as high as 1 in 10^7 [73] as e.g. shown for circulating tumor cells. Compared to IHC, there are more markers (i.e. mRNA transcripts) available that are highly expressed by malignant cells but not or only marginally present in the lymphatic tissue, which can lead to higher sensitivity of the assay. Also, the combined use of such tumor-associated markers can significantly improve assay specificity. For the investigation of LNs or BM preparations molecular detection has been successfully used in upper digestive tract cancers targeting mRNAs for cytokeratins, CEA, E48 (Ly-6D), TACSTD1 and Pemphigus vulgaris antigen (PVA) [74]. Another advantage of qRT-PCR is the possibility of automation and rapid processing, which might even permit intra-operative analysis. This method is very interesting for the clinical application to improve detection of micrometastasis or involved resection margins. However, due to the limited number of analyzed LNs, studies using this approach provide only limited information on the real extent of MRD since it is difficult to quantify an actual cell number based on the qPCR results. Because this is better possible with morphologic inspection, we focused our chapter only on DTCs visualized by immunostaining.

Similarly, qRT-PCR assays for the detection of DTCs in BM have been established. Most studies in upper digestive tract cancer investigate mRNA extracted from MNC fractions enriched by density centrifugation. CK18, CK19 and/or CEA mRNA were commonly used for molecular DTC detection in BM of esophageal cancer patients.

4.8 Clinical Relevance of MRD/DTCs in HNSCC and EC

LN metastases are the most important prognostic factor of upper digestive tract cancers and indicate a higher risk for systemic progression [22]. Yet, around 17% of the HNSCC patients classified as node-negative by routine histopathology (pN0) will develop recurrent disease in the first 5 years upon primary treatment despite initial successful surgery, indicating that relevant cancer spread escaped the detection level of clinical imaging and routine histopathology [1]. Most studies trying to trace this clinical occult tumor spread therefore aim at assessing the real extent of lymphatic cancer dissemination and at testing whether this is of prognostic significance. Thus, in case of lymphatic dissemination, the value of information from such studies is directly dependent on the extent of the lymphadenectomy and/or the number of analyzed LNs. Such data can inform on the necessary degree of LN dissection (e.g. whether the sentinel LN concept is feasible) or determine whether additional systemic therapies should be considered.

This information can be complemented by analyzing the BM for clinically occult cancer spread. Together with immunohistochemically detected DTCs in LNs, such analysis can provide a more realistic (yet incomplete) view on cancer cell spread, helping to explain metastatic relapses after initially successful surgery in M0 cancer patients.

4.9 LN-DTCs

There is a considerable amount of data on immunohistochemical detection of clinically occult LN-DTCs for HNSCC as well as EC. Since most studies used similar immunoassay protocols, antibody-clones, and screened a comparable number of LNs across both entities, the studies are roughly comparable (Table 4.1). Generally, the median detection rates of DTC and micrometastases (MMs) were quite similar between HNSCC (24%) and EC (27%) patients who have been diagnosed as pN0 in routine histopathology. According to these studies, almost one third of pN0-staged patients harbor undetected tumor cells in their LNs. Since even small primary tumors seeded cells into LNs, this indicates that tumor cell dissemination is a frequent and early event (Table 4.1). Two recent studies in EC, focusing on early cancers (pT1) in which the tumor reaches just either the lamina propria (M2), the muscularis mucosae (M3), or different depths of the submucosal layer (SM1-3), meticulously investigated this early dissemination. While in both reports, no lymphatic DTCs or MMs were detected in M2 tumors, one report detected DTCs in 8% of M3-tumors and even 25% of SM1-tumors [75]. This was corroborated by the second study revealing IHC positive tumor cells in the LNs of up to 50% of the cases with SM3 tumors [76]. This early dissemination in node-negative pT1 tumors is reflected by the occurrence of overt LN metastasis, requiring distinct surgical treatment. Since around 54% of patients with SM3 tumors present with already established LN metastases, submucosal tumors (SM1-3) require surgical resection with lymphadenectomy. In contrast, most mucosal (M1-3) tumors, in which almost no dissemination could be detected by IHC and in which LN metastasis were observed in just 2 cases, can safely be treated by endoscopic resection with excellent long-term results. These data further indicate that relevant dissemination rarely starts in tumors not invading the lamina propria. The mechanisms behind the early lymphatic dissemination of HNSCC and EC are still obscure and less well studied, if compared e.g. to breast cancer [77, 78]. However, epithelial to mesenchymal transition (EMT) may play a role in both cancers [79, 80] Transformed esophageal and oral keratinocytes, cells with clear epithelial properties, can become converted to cells with mesenchymal features in response to treatment with transforming growth factor-β (TGF-β), a potent inducer of EMT present in the tumor microenvironment [81–83]. Likewise, EMT seems to be highly relevant for progression of Barrett's esophagus [84], the precursor lesion for EAC. Several cues from inflamed or hypoxic microenvironments of the oral cavity or the esophagus can be potent triggers of EMT programs. Therefore, it is tempting to speculate that during early HNSCC/EC carcinogenesis EMT leads to dissemination of metastatic precursors. Once tumor cell dissemination has taken place, only few cells will be successful in establishing a lymphatic metastasis. Here, EpCAM appears to play a role governing a proliferative program and reverting the cells to an epithelial phenotype [85, 86]. The topographical distribution of DTCs follows that of established lymphatic metastases, targeting first the LNs closest to the primary tumor. In HNSCC, this has been meticulously studied in precisely mapped LNs, demonstrating a non-random distribution of DTCs with initial involvement of LNs most proximal to the primary tumor, followed by ipsilateral distribution with increasing distance to the primary tumor, and eventually contralateral involvement [87]. Although LN skipping (i.e. tumor cells/metastasis bypassing the orderly progression from one LN station/level to a contiguous LN level) was observed in both entities [87, 88], the early lymphatic dissemination in EC appears to be more erratic [89]. This might be explained by the very complex lymphatic anatomy of the esophagus, in which LNs do not drain segmentally, but rather in a longitudinal fashion, depending on the invasion depth of the tumor [36]. In addition, extensive communications between the esophageal layers are present. Due to the two different embryologic origins of the esophagus, esophageal draining LNs are also widely distributed with two main lymphatic collection areas, in the abdomen and at the neck. All of this can lead to the observed more erratic dissemination pattern in EC [38]. This is clinically relevant, since it argues against the application of sentinel LN biopsies in EC. However, no prognostic influence of extensive LN skipping was observed.

With regard to the prognostic relevance of DTCs and MMs, there is a striking difference between the two entities. While most studies observed a significant negative prognostic impact of DTCs and MMs in EC, so far, no study could find this in HNSCC. This is quite puzzling, considering the similar strong prognostic impact of overt LN metastases in both entities. One hypothesis is that in comprehensive neck-dissections in HNSCC, the lymphatic resection is more complete than in lymphadenectomies for esophageal cancer and thus relevant minimal residual disease is left behind. This is supported by randomized clinical trials in cN0-HNSCC-patients comparing the prognostic impact of elective neck dissections versus observation only, in which the incidence of delayed LN-metastases was up to six times higher in the observation only group and was also associated with a significantly shorter disease-free- and overall-survival [33]. As a consequence of the current HNSCC DTC data, an initial limited ipsilateral neck dissection with subsequent comprehensive immunohistochemical LN evaluation is warranted to decide on the extent of lymphadenectomy. From a biologic point of view, the absence of any prognostic impact of DTCs in HNSCC, even in studies with relatively long follow-up, suggests that lymphatic dissemination might not be an indicator of clinically relevant systemic dissemination. Apparently, most disseminated cells from early tumor stages (UICC I/II) do not have the capacity to grow into metastases outside of the lymphatic system during the HNSCC-patient's limited lifetime. Indirect evidence that possibly not all micrometastases and DTCs grow into macrometastases came from studies on a wait & see strategy in the cN0 neck in HNSCC, where the rate of conversion to a clinically positive neck (cN+) is substantially lower than the content of MMs and DTCs reported for elective ND specimens [90].

4.10 BM-DTCs

Patients with potentially curable HNSCC/EC are still at risk of developing metastatic relapses month or years after a complete tumor resection. Unfortunately, for such patients with early stage cancers and no LN metastasis, no clinical markers have been established to determine their individual risk. Addressing this problem, several studies have investigated the BM for DTCs as an indicator organ for clinical occult systemic cancer to elucidate whether BM DTCs could indicate a higher metastatic risk. In contrast to circulating tumor cells (CTC) detectable in peripheral blood that are shed from the primary tumor site with uncertain fate, BM DTCs potentially represent cancer cells already settled in premetastatic niches at a distant site. In the 28 studies investigating the prevalence of DTC in the BM aspirates of upper digestive tract cancer patients, the mean detection rate was 36% (range 7–88%) and thus higher compared to the detection rate of DTCs in LNs. The median DTC detection rate was slightly higher in EC (41%) than in HNSCC (34%), reflecting the overall higher incidence and more rapid development of distant metastasis in EC compared to HNSCC. However, it is important to note that the detected DTC numbers in BM aspirations of around 10 mL cannot be representative for the 1.75 L functional BM of an adult [91] and likely underestimates the extent of occult disease. This notion is supported by studies in EC that investigated the BM of ribs resected during transthoracic esophagectomy (Table), which revealed a significantly higher concentration of DTCs. The reason for this is unknown but was attributed to a better BM quality with a lesser proportion of fatty tissue and the avoidance of suction artifacts that may arise during the conventional iliac crest aspiration [67]. While iliac crest aspirations usually yielded 1–5 DTC per 10^6 BM cells, the lowest DTC-count in one of the rib BM studies was 20 DTC per 10^5 BM cells. Based on the average cellularity of human marrow of 11.1×10^9 nucleated cells/kg, the authors estimated the clinically occult residual systemic cancer in the BM to be in the order of 1.5×10^8 cancer cells for a 70 kg human. This suggests a substantial burden of minimal residual cancer after tumor resection, especially since it is likely that a comparable DTC burden exists in other organs that cannot be probed for DTCs, such as lungs and liver. Interestingly, in the majority of the HNSCC and EC studies, presence of BM DTCs conferred a negative prognostic impact in HNSCC and

EC. Despite this, the metastatic capacity of individual BM DTCs remains unclear. Since clinically relevant bone metastasis are very rarely detected in the lifetime of HNSCC or EC patients, it is tempting to speculate that further metastatic progression of DTCs is delayed by the dormancy inducing BM microenvironment [2, 49, 92, 93]. This hypothesis is supported by data from genomic analyses of single DTCs in esophageal cancer, revealing significantly lower aberration levels of DTCs isolated from BM when compared to DTCs isolated from LNs or to cancer tissue of matched primary tumors. This suggests dissemination at an early time point during tumor development when less mutations were present in the primary tumor. Once arrived in BM niches, these cells became locked in a dormant, non-proliferative state and failed to acquire new mutations due to the lack of cell division. The DTCs in LNs either proliferated, accumulated aberrations in the process, or were outcompeted by later arriving, more proliferative clones, explaining the higher aberration levels detected in this compartment [2, 49, 92, 93]. Yet, highly aberrant BM-DTCs and LN-DTCs were also detected along DTCs with low mutational load, and these were significantly correlated with poor survival, indicating that dissemination is a process that occurs constantly, beginning at an early time point.

4.11 Summary and Outlook

Immunohistochemical studies investigating DTCs in LNs and BM indicate that a substantial number of these cells are left behind after complete surgical resection in HNSCC and EC patients. These cells remain viable as MRD. Likely, these cells represent the potential metastatic progenitor cells, responsible for metastatic relapses. However, it is important to note that due to technical and economic restrictions, so far only the tip of the iceberg became detectable. All studies looking for LN-DTCs screened only a limited number of slides per LN or, in case of BM analysis, only a part of a limited sample (0.005% of the whole BM volume). Furthermore, the absence of tumor-specific markers as well as the unavailability of samples for such analysis render other potential metastatic sites such as lung or liver impossible to screen for minimal residual disease outside of animal experiments [94]. However, despite these obvious shortcomings, the data obtained on DTCs so far is of clinical and biological relevance. In HNSCC, live screening for DTCs in selected LNs close to the primary tumor could help to determine the required extend of lymphadenectomy; from a biologic point of view, BM DTCs in HNSCC as well as EC appear as very interesting targets for direct analysis to gain deeper insights into tumor cell dormancy. In addition, data from comprehensive genomic single cell analysis of DTCs did not only reveal genetic heterogeneity between in minimal residual disease [60, 95], but was promising for prognostic stratification and prediction in the clinical setting [60, 96]. However, aspiration of BM is quite uncomfortable for patients and repeated samplings are not practical. In contrast, the analysis of circulating tumor cells (CTC) in peripheral blood may have a higher clinical applicability and may allow real-time monitoring of the systemic disease. In the past 10 years significant technical progress has been made to enrich and isolate CTCs by molecular and/or biophysical methods. Similar to DTCs, the concentration of CTCs in the peripheral blood is quite low, especially in the MRD situation [97]. A potential solution would be a substantial increase of investigated blood volume to the scale of liters. This has become possible with the diagnostic leukapheresis (DLA) approach, as formally proven in a prospective study that analyzed side-by-side 7.5 mL of peripheral blood and 2 mL of an autologous DLA product. DLA led to a 2.5-fold increase in detection frequency and escalated the median CTC numbers 30-fold when 2 mL of the DLA product were analyzed [98]. We expect that in combination with different specific CTC-enrichment protocols, DLA will become an invaluable tool for in-depth analysis of minimal residual cancer in clinical as well as research applications.

References

1. NCI (2018) SEER Cancer Stat Facts: Oral Cavity and Pharynx Cancer 2018
2. Sosa MS, Bragado P, Aguirre-Ghiso JA (2014) Mechanisms of disseminated cancer cell dormancy: an awakening field. Nat Rev Cancer 14(9):611–622. https://doi.org/10.1038/nrc3793
3. NCI (2018) SEER Cancer Stat Facts: Esophageal Cancer 2018
4. Ai D, Zhu H, Ren W, Chen Y, Liu Q, Deng J, Ye J, Fan J, Zhao K (2017) Patterns of distant organ metastases in esophageal cancer: a population-based study. J Thorac Dis 9(9):3023–3030. https://doi.org/10.21037/jtd.2017.08.72
5. Nakamura Y, Togashi Y, Nakahara H, Tomida S, Banno E, Terashima M, Hayashi H, de Velasco MA, Sakai K, Fujita Y, Okegawa T, Nutahara K, Hamada S, Nishio K (2016) Afatinib against esophageal or head-and-neck squamous cell carcinoma: Significance of activating oncogenic HER4 mutations in HNSCC. Mol Cancer Ther 15(8):1988–1997. https://doi.org/10.1158/1535-7163.MCT-15-0737
6. Califano J, van der Riet P, Westra W, Nawroz H, Clayman G, Piantadosi S, Corio R, Lee D, Greenberg B, Koch W, Sidransky D (1996) Genetic progression model for head and neck cancer: implications for field cancerization. Cancer Res 56(11):2488–2492
7. Perez-Ordonez B, Beauchemin M, Jordan RC (2006) Molecular biology of squamous cell carcinoma of the head and neck. J Clin Pathol 59(5):445–453. https://doi.org/10.1136/jcp.2003.007641
8. Leemans CR, Braakhuis BJ, Brakenhoff RH (2011) The molecular biology of head and neck cancer. Nat Rev Cancer 11(1):9–22. https://doi.org/10.1038/nrc2982
9. Pai SI, Westra WH (2009) Molecular pathology of head and neck cancer: implications for diagnosis, prognosis, and treatment. Annu Rev Pathol 4:49–70. https://doi.org/10.1146/annurev.pathol.4.110807.092158
10. Klein CA, Stoecklein NH (2009) Lessons from an aggressive cancer: evolutionary dynamics in esophageal carcinoma. Cancer Res 69(13):5285–5288. https://doi.org/10.1158/0008-5472.CAN-08-4586
11. Boje CR (2014) Impact of comorbidity on treatment outcome in head and neck squamous cell carcinoma – a systematic review. Radiother Oncol 110(1):81–90. https://doi.org/10.1016/j.radonc.2013.07.005
12. Adrien J, Bertolus C, Gambotti L, Mallet A, Baujat B (2014) Why are head and neck squamous cell carcinoma diagnosed so late? Influence of health care disparities and socio-economic factors. Oral Oncol 50(2):90–97. https://doi.org/10.1016/j.oraloncology.2013.10.016
13. Simpson MC, Massa ST, Boakye EA, Antisdel JL, Stamatakis KA, Varvares MA, Osazuwa-Peters N (2018) Primary cancer vs competing causes of death in survivors of head and neck cancer. JAMA Oncol 4(2):257–259. https://doi.org/10.1001/jamaoncol.2017.4478
14. Simard EP, Ward EM, Siegel R, Jemal A (2012) Cancers with increasing incidence trends in the United States: 1999 through 2008. CA Cancer J Clin 62(2):118–128. https://doi.org/10.3322/caac.20141
15. Saeed NA, Mellon EA, Meredith KL, Hoffe SE, Shridhar R, Frakes J, Fontaine JP, Pimiento JM, Kothari N, Almhanna K (2017) Adjuvant chemotherapy and outcomes in esophageal carcinoma. J Gastrointest Oncol 8(5):816–824. https://doi.org/10.21037/jgo.2017.07.10
16. Mallen-St Clair J, Alani M, Wang MB, Srivatsan ES (2016) Human papillomavirus in oropharyngeal cancer: the changing face of a disease. Biochim Biophys Acta 1866(2):141–150. https://doi.org/10.1016/j.bbcan.2016.07.005
17. Nygard M, Aagnes B, Bray F, Moller B, Mork J (2012) Population-based evidence of increased survival in human papillomavirus-related head and neck cancer. Eur J Cancer 48(9):1341–1346. https://doi.org/10.1016/j.ejca.2012.03.014
18. Shen W, Sakamoto N, Yang L (2015) Cancer-specific mortality and competing mortality in patients with head and neck squamous cell carcinoma: a competing risk analysis. Ann Surg Oncol 22(1):264–271. https://doi.org/10.1245/s10434-014-3951-8
19. Kulasinghe A, Perry C, Jovanovic L, Nelson C, Punyadeera C (2015) Circulating tumour cells in metastatic head and neck cancers. Int J Cancer 136(11):2515–2523. https://doi.org/10.1002/ijc.29108
20. Vallbohmer D, Holscher AH, DeMeester S, DeMeester T, Salo J, Peters J, Lerut T, Swisher SG, Schroder W, Bollschweiler E, Hofstetter W (2010) A multicenter study of survival after neoadjuvant radiotherapy/chemotherapy and esophagectomy for ypT0N0M0R0 esophageal cancer. Ann Surg 252(5):744–749. https://doi.org/10.1097/SLA.0b013e3181fb8dde
21. Humm JL, Rosenfeld A, Del Guerra A (2003) From PET detectors to PET scanners. Eur J Nucl Med Mol Imaging 30(11):1574–1597. https://doi.org/10.1007/s00259-003-1266-2
22. Woolgar JA, Triantafyllou A, Lewis JS Jr, Hunt J, Williams MD, Takes RP, Thompson LD, Slootweg PJ, Devaney KO, Ferlito A (2013) Prognostic biological features in neck dissection specimens. Eur Arch Otorhinolaryngol 270(5):1581–1592. https://doi.org/10.1007/s00405-012-2170-9
23. Rouviére H (1932) Anatomie des lymphatiques de l'homme Masson, Paris
24. Rouviére H (1938) Lymphatic system of the head and neck. In: (Translator) Tobias M (ed) Anatomy of the human lymphatic system. Edwards, Ann Arbor
25. Werner JA, Davis RK (2004) Metastases in head and neck cancer. Springer, New York
26. Shah JP, Strong E, Spiro RH, Vikram B (1981) Surgical grand rounds. Neck dissection: current status and future possibilities. Clin Bull 11(1):25–33

27. Shah JP (1990) Patterns of cervical lymph node metastasis from squamous carcinomas of the upper aerodigestive tract. Am J Surg 160(4):405–409
28. Robbins KT, Clayman G, Levine PA, Medina J, Sessions R, Shaha A, Som P, Wolf GT, American H, Neck S, American Academy of O-H, Neck S (2002) Neck dissection classification update: revisions proposed by the American Head and Neck Society and the American Academy of Otolaryngology-Head and Neck Surgery. Arch Otolaryngol Head Neck Surg 128(7):751–758
29. Amin MB, Edge SB, Greene FL, Byrd DR, Brookland RK, Washington MK, Gershenwald JE, Compton CC, Hess KR, Sullivan DC, Jessup JM, Brierley JD, Gaspar LE, Schilsky RL, Balch CM, Winchester DP, Asare EA, Madera M, Gress DM, Meyer LRE (2017) AJCC cancer staging manual, 8th edn. Springer, New York
30. Crile GW (1923) Carcinoma of the jaw, tongue, cheek, and lips. Surg Gynecol Obstet 36:159–184
31. Medina JE (1989) A rational classification of neck dissections. Otolaryngol Head Neck Surg 100(3):169–176. https://doi.org/10.1177/019459988910000301
32. Brazilian Head and Neck Cancer Study Group (1998) Results of a prospective trial on elective modified radical classical versus supraomohyoid neck dissection in the management of oral squamous carcinoma. Am J Surg 176(5):422–427
33. D'Cruz AK, Vaish R, Kapre N, Dandekar M, Gupta S, Hawaldar R, Agarwal JP, Pantvaidya G, Chaukar D, Deshmukh A, Kane S, Arya S, Ghosh-Laskar S, Chaturvedi P, Pai P, Nair S, Nair D, Badwe R, Head, Neck Disease Management G (2015) Elective versus therapeutic neck dissection in node-negative oral cancer. N Engl J Med 373(6):521–529. https://doi.org/10.1056/NEJMoa1506007
34. Peyre CG, Hagen JA, DeMeester SR, Van Lanschot JJ, Holscher A, Law S, Ruol A, Ancona E, Griffin SM, Altorki NK, Rice TW, Wong J, Lerut T, DeMeester TR (2008) Predicting systemic disease in patients with esophageal cancer after esophagectomy: a multinational study on the significance of the number of involved lymph nodes. Ann Surg 248(6):979–985. https://doi.org/10.1097/SLA.0b013e3181904f3c
35. Kuge K, Murakami G, Mizobuchi S, Hata Y, Aikou T, Sasaguri S (2003) Submucosal territory of the direct lymphatic drainage system to the thoracic duct in the human esophagus. J Thorac Cardiovasc Surg 125(6):1343–1349
36. Tachimori Y (2017) Pattern of lymph node metastases of squamous cell esophageal cancer based on the anatomical lymphatic drainage system: efficacy of lymph node dissection according to tumor location. J Thorac Dis 9(Suppl 8):S724–S730. https://doi.org/10.21037/jtd.2017.06.19
37. Society JE (2009) Japanese classification of esophageal cancer, vol 6, Part 1: Esophagus, 10th edn. Kanehara, Tokyo
38. Matsuda S, Takeuchi H, Kawakubo H, Kitagawa Y (2017) Three-field lymph node dissection in esophageal cancer surgery. J Thorac Dis 9(Suppl 8):S731–S740. https://doi.org/10.21037/jtd.2017.03.171
39. Rice TW, Ishwaran H, Hofstetter WL, Schipper PH, Kesler KA, Law S, Lerut EM, Denlinger CE, Salo JA, Scott WJ, Watson TJ, Allen MS, Chen LQ, Rusch VW, Cerfolio RJ, Luketich JD, Duranceau A, Darling GE, Pera M, Apperson-Hansen C, Blackstone EH (2017) Esophageal cancer: associations with (pN+) lymph node metastases. Ann Surg 265(1):122–129. https://doi.org/10.1097/SLA.0000000000001594
40. Tachimori Y, Ozawa S, Numasaki H, Matsubara H, Shinoda M, Toh Y, Udagawa H, Fujishiro M, Oyama T, Uno T, Registration Committee for Esophageal Cancer of the Japan Esophageal S (2016) Efficacy of lymph node dissection by node zones according to tumor location for esophageal squamous cell carcinoma. Esophagus 13:1–7. https://doi.org/10.1007/s10388-015-0515-3
41. Rizk NP, Ishwaran H, Rice TW, Chen LQ, Schipper PH, Kesler KA, Law S, Lerut TE, Reed CE, Salo JA, Scott WJ, Hofstetter WL, Watson TJ, Allen MS, Rusch VW, Blackstone EH (2010) Optimum lymphadenectomy for esophageal cancer. Ann Surg 251(1):46–50. https://doi.org/10.1097/SLA.0b013e3181b2f6ee
42. Takeuchi H, Fujii H, Ando N, Ozawa S, Saikawa Y, Suda K, Oyama T, Mukai M, Nakahara T, Kubo A, Kitajima M, Kitagawa Y (2009) Validation study of radio-guided sentinel lymph node navigation in esophageal cancer. Ann Surg 249(5):757–763. https://doi.org/10.1097/SLA.0b013e3181a38e89
43. Ma GW, Situ DR, Ma QL, Long H, Zhang LJ, Lin P, Rong TH (2014) Three-field vs two-field lymph node dissection for esophageal cancer: a meta-analysis. World J Gastroenterol 20(47):18022–18030. https://doi.org/10.3748/wjg.v20.i47.18022
44. Topazian DS (1961) Distant metastasis of oral carcinoma. Oral Surg Oral Med Oral Pathol 14:705–711
45. Probert JC, Thompson RW, Bagshaw MA (1974) Patterns of spread of distant metastases in head and neck cancer. Cancer 33(1):127–133
46. Papac RJ (1984) Distant metastases from head and neck cancer. Cancer 53(2):342–345
47. Hsu L-P, Chen P-R (2005) Distant metastases of head and neck squamous cell carcinomas – experience from Eastern Taiwan. Tzu Chi Med J 17:99–104
48. Riethdorf S, Pantel K (2009) Clinical relevance and current challenges of research on disseminating tumor cells in cancer patients. Breast Cancer Res 11(Suppl 3):S10. https://doi.org/10.1186/bcr2429
49. Pantel K, Alix-Panabieres C (2014) Bone marrow as a reservoir for disseminated tumor cells: a special source for liquid biopsy in cancer patients. Bonekey Rep 3:584. https://doi.org/10.1038/bonekey.2014.79
50. Sobin LH, Wittekind C (2014) TNM classification of malignant tumours, 6th edn. Wiley, Hoboken
51. Bragado P, Estrada Y, Parikh F, Krause S, Capobianco C, Farina HG, Schewe DM, Aguirre-Ghiso JA (2013) TGF-beta2 dictates disseminated tumour cell fate in target organs through TGF-beta-RIII and p38alpha/

beta signalling. Nat Cell Biol 15(11):1351–1361. https://doi.org/10.1038/ncb2861
52. Hermanek P, Hutter RV, Sobin LH, Wittekind C (1999) International union against cancer. classification of isolated tumor cells and micrometastasis. Cancer 86(12):2668–2673
53. Woolgar JA, Triantafyllou A (2011) Squamous cell carcinoma and precursor lesions: clinical pathology. Periodontol 2000 57(1):51–72. https://doi.org/10.1111/j.1600-0757.2011.00389.x
54. Lewis JS Jr, Carpenter DH, Thorstad WL, Zhang Q, Haughey BH (2011) Extracapsular extension is a poor predictor of disease recurrence in surgically treated oropharyngeal squamous cell carcinoma. Mod Pathol 24(11):1413–1420. https://doi.org/10.1038/modpathol.2011.105
55. Ambrosch P, Kron M, Fischer G, Brinck U (1995) Micrometastases in carcinoma of the upper aerodigestive tract: detection, risk of metastasizing, and prognostic value of depth of invasion. Head Neck 17(6):473–479
56. Broglie MA, Haerle SK, Huber GF, Haile SR, Stoeckli SJ (2013) Occult metastases detected by sentinel node biopsy in patients with early oral and oropharyngeal squamous cell carcinomas: impact on survival. Head Neck 35(5):660–666. https://doi.org/10.1002/hed.23017
57. Garrel R, Dromard M, Costes V, Barbotte E, Comte F, Gardiner Q, Cartier C, Makeieff M, Crampette L, Guerrier B, Boulle N (2006) The diagnostic accuracy of reverse transcription-PCR quantification of cytokeratin mRNA in the detection of sentinel lymph node invasion in oral and oropharyngeal squamous cell carcinoma: a comparison with immunohistochemistry. Clin Cancer Res 12(8):2498–2505. https://doi.org/10.1158/1078-0432.CCR-05-2136
58. Solassol J, Burcia V, Costes V, Lacombe J, Mange A, Barbotte E, de Verbizier D, Cartier C, Makeieff M, Crampette L, Boulle N, Maudelonde T, Guerrier B, Garrel R (2010) Pemphigus vulgaris antigen mRNA quantification for the staging of sentinel lymph nodes in head and neck cancer. Br J Cancer 102(1):181–187. https://doi.org/10.1038/sj.bjc.6605470
59. Ulmer A, Fischer JR, Schanz S, Sotlar K, Breuninger H, Dietz K, Fierlbeck G, Klein CA (2005) Detection of melanoma cells displaying multiple genomic changes in histopathologically negative sentinel lymph nodes. Clin Cancer Res 11(15):5425–5432. https://doi.org/10.1158/1078-0432.CCR-04-1995
60. Stoecklein NH, Hosch SB, Bezler M, Stern F, Hartmann CH, Vay C, Siegmund A, Scheunemann P, Schurr P, Knoefel WT, Verde PE, Reichelt U, Erbersdobler A, Grau R, Ullrich A, Izbicki JR, Klein CA (2008) Direct genetic analysis of single disseminated cancer cells for prediction of outcome and therapy selection in esophageal cancer. Cancer Cell 13(5):441–453. https://doi.org/10.1016/j.ccr.2008.04.005
61. Karantza V (2011) Keratins in health and cancer: more than mere epithelial cell markers. Oncogene 30(2):127–138. https://doi.org/10.1038/onc.2010.456
62. Gould VE, Bloom KJ, Franke WW, Warren WH, Moll R (1995) Increased numbers of cytokeratin-positive interstitial reticulum cells (CIRC) in reactive, inflammatory and neoplastic lymphadenopathies: hyperplasia or induced expression? Virchows Arch 425(6):617–629
63. Xu X, Roberts SA, Pasha TL, Zhang PJ (2000) Undesirable cytokeratin immunoreactivity of native nonepithelial cells in sentinel lymph nodes from patients with breast carcinoma. Arch Pathol Lab Med 124(9):1310–1313. https://doi.org/10.1043/0003-9985(2000)124<1310:UCIONN>2.0.CO;2
64. Ribourtout B, Zandecki M (2015) Plasma cell morphology in multiple myeloma and related disorders. Morphologie 99(325):38–62. https://doi.org/10.1016/j.morpho.2015.02.001
65. Brown RB, Gaillard RA, Turner JA (1953) The significance of aberrant or heterotopic parotid gland tissue in lymph nodes. Ann Surg 138(6):850–856
66. Passlick B, Izbicki JR, Kubuschok B, Nathrath W, Thetter O, Pichlmeier U, Schweiberer L, Riethmuller G, Pantel K (1994) Immunohistochemical assessment of individual tumor cells in lymph nodes of patients with non-small-cell lung cancer. J Clin Oncol 12(9):1827–1832. https://doi.org/10.1200/JCO.1994.12.9.1827
67. O'Sullivan GC, Sheehan D, Clarke A, Stuart R, Kelly J, Kiely MD, Walsh T, Collins JK, Shanahan F (1999) Micrometastases in esophagogastric cancer: high detection rate in resected rib segments. Gastroenterology 116(3):543–548
68. Noguchi T, Shibata T, Fumoto S, Sato T, Uchida Y, Daa T, Yokoyama S, Gabbert HE, Mueller W, Takeno S (2003) Detection of disseminated cancer cells in rib marrow of patients with esophageal cancer. Oncol Rep 10(3):623–627
69. Lin H, Balic M, Zheng S, Datar R, Cote RJ (2011) Disseminated and circulating tumor cells: role in effective cancer management. Crit Rev Oncol Hematol 77(1):1–11. https://doi.org/10.1016/j.critrevonc.2010.04.008
70. Vincent-Salomon A, Bidard FC, Pierga JY (2008) Bone marrow micrometastasis in breast cancer: review of detection methods, prognostic impact and biological issues. J Clin Pathol 61(5):570–576. https://doi.org/10.1136/jcp.2007.046649
71. Klein CA, Schmidt-Kittler O, Schardt JA, Pantel K, Speicher MR, Riethmuller G (1999) Comparative genomic hybridization, loss of heterozygosity, and DNA sequence analysis of single cells. Proc Natl Acad Sci U S A 96(8):4494–4499
72. Borgen E, Pantel K, Schlimok G, Muller P, Otte M, Renolen A, Ehnle S, Coith C, Nesland JM, Naume B (2006) A European interlaboratory testing of three well-known procedures for immunocytochemical detection of epithelial cells in bone marrow. Results from analysis of normal bone marrow. Cytometry B

Clin Cytom 70(6):400–409. https://doi.org/10.1002/cyto.b.20114
73. Mori M, Mimori K, Ueo H, Karimine N, Barnard GF, Sugimachi K, Akiyoshi T (1996) Molecular detection of circulating solid carcinoma cells in the peripheral blood: the concept of early systemic disease. Int J Cancer 68(6):739–743. https://doi.org/10.1002/(SICI)1097-0215(19961211)68:6<739::AID-IJC8>3.0.CO;2-4
74. Nieuwenhuis EJ, Leemans CR, Kummer JA, Denkers F, Snow GB, Brakenhoff RH (2003) Assessment and clinical significance of micrometastases in lymph nodes of head and neck cancer patients detected by E48 (Ly-6D) quantitative reverse transcription-polymerase chain reaction. Lab Investig 83(8):1233–1240
75. Grotenhuis BA, van Heijl M, Wijnhoven BP, van Berge Henegouwen MI, Biermann K, ten Kate FJ, Busch OR, Dinjens WN, Tilanus HW, van Lanschot JJ (2010) Lymphatic micrometastases in patients with early esophageal adenocarcinoma. J Surg Oncol 102(7):863–867. https://doi.org/10.1002/jso.21719
76. Prenzel KL, Holscher AH, Drebber U, Agavonova M, Gutschow CA, Bollschweiler E (2012) Prognostic impact of nodal micrometastasis in early esophageal cancer. Eur J Surg Oncol 38(4):314–318. https://doi.org/10.1016/j.ejso.2012.01.007
77. Harper KL, Sosa MS, Entenberg D, Hosseini H, Cheung JF, Nobre R, Avivar-Valderas A, Nagi C, Girnius N, Davis RJ, Farias EF, Condeelis J, Klein CA, Aguirre-Ghiso JA (2016) Mechanism of early dissemination and metastasis in Her2(+) mammary cancer. Nature 540:588–592. https://doi.org/10.1038/nature20609
78. Hosseini H, Obradovic MM, Hoffmann M, Harper KL, Sosa MS, Werner-Klein M, Nanduri LK, Werno C, Ehrl C, Maneck M, Patwary N, Haunschild G, Guzvic M, Reimelt C, Grauvogl M, Eichner N, Weber F, Hartkopf AD, Taran FA, Brucker SY, Fehm T, Rack B, Buchholz S, Spang R, Meister G, Aguirre-Ghiso JA, Klein CA (2016) Early dissemination seeds metastasis in breast cancer. Nature 540:552–558. https://doi.org/10.1038/nature20785
79. Smith A, Teknos TN, Pan Q (2013) Epithelial to mesenchymal transition in head and neck squamous cell carcinoma. Oral Oncol 49(4):287–292. https://doi.org/10.1016/j.oraloncology.2012.10.009
80. Okabe H, Mima K, Saito S, Hayashi H, Imai K, Nitta H, Hashimoto D, Chikamoto A, Ishiko T, Beppu T, Baba H (2015) Epithelial-mesenchymal transition in gastroenterological cancer. J Cancer Metastasis Treat 1:183–189
81. Whelan KA, Chandramouleeswaran PM, Tanaka K, Natsuizaka M, Guha M, Srinivasan S, Darling DS, Kita Y, Natsugoe S, Winkler JD, Klein-Szanto AJ, Amaravadi RK, Avadhani NG, Rustgi AK, Nakagawa H (2017) Autophagy supports generation of cells with high CD44 expression via modulation of oxidative stress and Parkin-mediated mitochondrial clearance. Oncogene 36(34):4843–4858. https://doi.org/10.1038/onc.2017.102
82. Kinugasa H, Whelan KA, Tanaka K, Natsuizaka M, Long A, Guo A, Chang S, Kagawa S, Srinivasan S, Guha M, Yamamoto K, St Clair DK, Avadhani NG, Diehl JA, Nakagawa H (2015) Mitochondrial SOD2 regulates epithelial-mesenchymal transition and cell populations defined by differential CD44 expression. Oncogene 34(41):5229–5239. https://doi.org/10.1038/onc.2014.449
83. Natsuizaka M, Whelan KA, Kagawa S, Tanaka K, Giroux V, Chandramouleeswaran PM, Long A, Sahu V, Darling DS, Que J, Yang Y, Katz JP, Wileyto EP, Basu D, Kita Y, Natsugoe S, Naganuma S, Klein-Szanto AJ, Diehl JA, Bass AJ, Wong KK, Rustgi AK, Nakagawa H (2017) Interplay between Notch1 and Notch3 promotes EMT and tumor initiation in squamous cell carcinoma. Nat Commun 8(1):1758. https://doi.org/10.1038/s41467-017-01500-9
84. Clark RJ, Craig MP, Agrawal S, Kadakia M (2018) microRNA involvement in the onset and progression of Barrett's esophagus: a systematic review. Oncotarget 9(8):8179–8196. https://doi.org/10.18632/oncotarget.24145
85. Driemel C, Kremling H, Schumacher S, Will D, Wolters J, Lindenlauf N, Mack B, Baldus SA, Hoya V, Pietsch JM, Panagiotidou P, Raba K, Vay C, Vallbohmer D, Harreus U, Knoefel WT, Stoecklein NH, Gires O (2014) Context-dependent adaption of EpCAM expression in early systemic esophageal cancer. Oncogene 33(41):4904–4915. https://doi.org/10.1038/onc.2013.441
86. Gires O, Stoecklein NH (2014) Dynamic EpCAM expression on circulating and disseminating tumor cells: causes and consequences. Cell Mol Life Sci 71(22):4393–4402. https://doi.org/10.1007/s00018-014-1693-1
87. Sproll C, Freund AK, Hassel A, Holbling M, Aust V, Storb SH, Handschel J, Teichmann C, Depprich R, Behrens B, Neves RP, Kubler NR, Kaiser P, Baldus SE, Toth C, Kaisers W, Stoecklein NH (2017) Immunohistochemical detection of lymph node-DTCs in patients with node-negative HNSCC. Int J Cancer 140(9):2112–2124. https://doi.org/10.1002/ijc.30617
88. Hosch SB, Stoecklein NH, Pichlmeier U, Rehders A, Scheunemann P, Niendorf A, Knoefel WT, Izbicki JR (2001) Esophageal cancer: the mode of lymphatic tumor cell spread and its prognostic significance. J Clin Oncol 19(7):1970–1975. https://doi.org/10.1200/JCO.2001.19.7.1970
89. Hosch S, Kraus J, Scheunemann P, Izbicki JR, Schneider C, Schumacher U, Witter K, Speicher MR, Pantel K (2000) Malignant potential and cytogenetic characteristics of occult disseminated tumor cells in esophageal cancer. Cancer Res 60(24):6836–6840
90. Pitman KT (2000) Rationale for elective neck dissection. Am J Otolaryngol 21(1):31–37

91. Hassan HT, El-Sheemy M (2004) Adult bone-marrow stem cells and their potential in medicine. J R Soc Med 97(10):465–471. https://doi.org/10.1258/jrsm.97.10.465
92. Aguirre-Ghiso JA, Bragado P, Sosa MS (2013) Metastasis awakening: targeting dormant cancer. Nat Med 19(3):276–277. https://doi.org/10.1038/nm.3120
93. Guzvic M, Klein CA (2013) Cancer dormancy: time to explore its clinical relevance. Breast Cancer Res 15(6):321. https://doi.org/10.1186/bcr3590
94. Fluegen G, Avivar-Valderas A, Wang Y, Padgen MR, Williams JK, Nobre AR, Calvo V, Cheung JF, Bravo-Cordero JJ, Entenberg D, Castracane J, Verkhusha V, Keely PJ, Condeelis J, Aguirre-Ghiso JA (2017) Phenotypic heterogeneity of disseminated tumour cells is preset by primary tumour hypoxic microenvironments. Nat Cell Biol 19(2):120–132. https://doi.org/10.1038/ncb3465
95. Schumacher S, Bartenhagen C, Hoffmann M, Will D, Fischer JC, Baldus SE, Vay C, Fluegen G, Dizdar L, Vallbohmer D, Klein CA, Knoefel WT, Stoecklein NH, Mohlendick B (2017) Disseminated tumour cells with highly aberrant genomes are linked to poor prognosis in operable oesophageal adenocarcinoma. Br J Cancer 117(5):725–733. https://doi.org/10.1038/bjc.2017.233
96. Hoffmann M, Pasch S, Schamberger T, Maneck M, Mohlendick B, Schumacher S, Brockhoff G, Knoefel WT, Izbicki J, Polzer B, Stoecklein NH, Klein CA (2018) Diagnostic pathology of early systemic cancer: ERBB2 gene amplification in single disseminated cancer cells determines patient survival in operable esophageal cancer. Int J Cancer 142(4):833–843. https://doi.org/10.1002/ijc.31108
97. Stoecklein NH, Fischer JC, Niederacher D, Terstappen LW (2016) Challenges for CTC-based liquid biopsies: low CTC frequency and diagnostic leukapheresis as a potential solution. Expert Rev Mol Diagn 16(2):147–164. https://doi.org/10.1586/14737159.2016.1123095
98. Fischer JC, Niederacher D, Topp SA, Honisch E, Schumacher S, Schmitz N, Zacarias Fohrding L, Vay C, Hoffmann I, Kasprowicz NS, Hepp PG, Mohrmann S, Nitz U, Stresemann A, Krahn T, Henze T, Griebsch E, Raba K, Rox JM, Wenzel F, Sproll C, Janni W, Fehm T, Klein CA, Knoefel WT, Stoecklein NH (2013) Diagnostic leukapheresis enables reliable detection of circulating tumor cells of nonmetastatic cancer patients. Proc Natl Acad Sci U S A 110(41):16580–16585. https://doi.org/10.1073/pnas.1313594110
99. Izbicki JR, Hosch SB, Pichlmeier U, Rehders A, Busch C, Niendorf A, Passlick B, Broelsch CE, Pantel K (1997) Prognostic value of immunohistochemically identifiable tumor cells in lymph nodes of patients with completely resected esophageal cancer. N Engl J Med 337(17):1188–1194. https://doi.org/10.1056/NEJM199710233371702
100. Natsugoe S, Mueller J, Stein HJ, Feith M, Hofler H, Siewert JR (1998) Micrometastasis and tumor cell microinvolvement of lymph nodes from esophageal squamous cell carcinoma: frequency, associated tumor characteristics, and impact on prognosis. Cancer 83(5):858–866
101. Glickman JN, Torres C, Wang HH, Turner JR, Shahsafaei A, Richards WG, Sugarbaker DJ, Odze RD (1999) The prognostic significance of lymph node micrometastasis in patients with esophageal carcinoma. Cancer 85(4):769–778
102. Matsumoto M, Natsugoe S, Nakashima S, Sakamoto F, Okumura H, Sakita H, Baba M, Takao S, Aikou T (2000) Clinical significance of lymph node micrometastasis of pN0 esophageal squamous cell carcinoma. Cancer Lett 153(1–2):189–197
103. Komukai S, Nishimaki T, Watanabe H, Ajioka Y, Suzuki T, Hatakeyama K (2000) Significance of immunohistochemically demonstrated micrometastases to lymph nodes in esophageal cancer with histologically negative nodes. Surgery 127(1):40–46. https://doi.org/10.1067/msy.2000.102754
104. Sato F, Shimada Y, Li Z, Watanabe G, Maeda M, Imamura M (2001) Lymph node micrometastasis and prognosis in patients with oesophageal squamous cell carcinoma. Br J Surg 88(3):426–432. https://doi.org/10.1046/j.1365-2168.2001.01687.x
105. Qubain SW, Natsugoe S, Matsumoto M, Nakashima S, Baba M, Takao S, Aikou T (2001) Micrometastases in the cervical lymph nodes in esophageal squamous cell carcinoma. Dis Esophagus 14(2):143–148
106. Komukai S, Nishimaki T, Suzuki T, Kanda T, Kuwabara S, Hatakeyama K (2002) Significance of immunohistochemical nodal micrometastasis as a prognostic indicator in potentially curable oesophageal carcinoma. Br J Surg 89(2):213–219. https://doi.org/10.1046/j.0007-1323.2001.01981.x
107. Doki Y, Ishikawa O, Mano M, Hiratsuka M, Sasaki Y, Kameyama M, Ohigashi H, Murata K, Yamada T, Miyashiro I, Yokoyama S, Ishiguro S, Imaoka S (2002) Cytokeratin deposits in lymph nodes show distinct clinical significance from lymph node micrometastasis in human esophageal cancers. J Surg Res 107(1):75–81
108. Vazquez-Sequeiros E, Wang L, Burgart L, Harmsen W, Zinsmeister A, Allen M, Jondal M, Wiersema M (2002) Occult lymph node metastases as a predictor of tumor relapse in patients with node-negative esophageal carcinoma. Gastroenterology 122(7):1815–1821
109. Waterman TA, Hagen JA, Peters JH, DeMeester SR, Taylor CR, Demeester TR (2004) The prognostic importance of immunohistochemically detected node metastases in resected esophageal adenocarcinoma. Ann Thorac Surg 78(4):1161–1169; discussion 1161–1169. https://doi.org/10.1016/j.athoracsur.2004.04.045
110. Heeren PA, Kelder W, Blondeel I, van Westreenen HL, Hollema H, Plukker JT (2005) Prognostic value of nodal micrometastases in patients with can-

cer of the gastro-oesophageal junction. Eur J Surg Oncol 31(3):270–276. https://doi.org/10.1016/j.ejso.2004.12.001

111. McGuill MJ, Byrne P, Ravi N, Reynolds J (2008) The prognostic impact of occult lymph node metastasis in cancer of the esophagus or esophago-gastric junction: systematic review and meta-analysis. Dis Esophagus 21(3):236–240. https://doi.org/10.1111/j.1442-2050.2007.00765.x
112. Buskens CJ, Ten Kate FJ, Obertop H, Izbicki JR, van Lanschot JJ (2008) Analysis of micrometastatic disease in histologically negative lymph nodes of patients with adenocarcinoma of the distal esophagus or gastric cardia. Dis Esophagus 21(6):488–495. https://doi.org/10.1111/j.1442-2050.2007.00805.x
113. Zingg U, Montani M, Busch M, Metzger U, Went P, Oertli D (2009) Prognostic influence of immunohistochemically detected lymph node micrometastasis and histological subtype in pN0 oesophageal cancer. Eur J Surg Oncol 35(6):593–599. https://doi.org/10.1016/j.ejso.2008.12.001
114. Koenig AM, Prenzel KL, Bogoevski D, Yekebas EF, Bubenheim M, Faithova L, Vashist YK, Gawad KA, Baldus SE, Pantel K, Schneider PM, Holscher AH, Izbicki JR (2009) Strong impact of micrometastatic tumor cell load in patients with esophageal carcinoma. Ann Surg Oncol 16(2):454–462. https://doi.org/10.1245/s10434-008-0169-7
115. Chao YK, Yeh CJ, Chuang WY, Fan KH, Hsieh MJ, Chu Y, Cheng AJ, Liu HP (2009) Prognostic significance of immunohistochemically detected lymph node micrometastases in pT0N0 esophageal squamous cell carcinoma. J Surg Oncol 100(7):559–562. https://doi.org/10.1002/jso.21362
116. Thompson SK, Ruszkiewicz AR, Jamieson GG, Sullivan TR, Devitt PG (2010) Isolated tumor cells in esophageal cancer: implications for the surgeon and the pathologist. Ann Surg 252(2):299–306. https://doi.org/10.1097/SLA.0b013e3181e61e15
117. Yanagi M, Sasaki K, Uchikado Y, Omoto I, Arigami T, Kurahara H, Uenosono Y, Sakoda M, Maemura K, Natsugoe S (2018) Effect of neoadjuvant chemoradiotherapy on lymph node micrometastases in thoracic esophageal cancer. Anticancer Res 38(2):893–900. https://doi.org/10.21873/anticanres.12299
118. Enepekides DJ, Sultanem K, Nguyen C, Shenouda G, Black MJ, Rochon L (1999) Occult cervical metastases: immunoperoxidase analysis of the pathologically negative neck. Otolaryngol Head Neck Surg 120(5):713–717. https://doi.org/10.1053/hn.1999.v120.a91761
119. Yu Z, Han D, Fan E (1999) Immunohistochemical evaluation of micrometastases to cervical lymph nodes in supraglottic cancer. Zhonghua Er Bi Yan Hou Ke Za Zhi 34(5):306–308
120. Kocaturk S, Okuyucu S, Koybasioglu F, Kacar A, Erkam U (2003) The importance of the Delphian lymph node in subtotal laryngeal surgery. Kulak Burun Bogaz Ihtis Derg 10(6):231–236
121. Thorban S, Roder JD, Nekarda H, Funk A, Siewert JR, Pantel K (1996) Immunocytochemical detection of disseminated tumor cells in the bone marrow of patients with esophageal carcinoma. J Natl Cancer Inst 88(17):1222–1227
122. Thorban S, Rosenberg R, Busch R, Roder RJ (2000) Epithelial cells in bone marrow of oesophageal cancer patients: a significant prognostic factor in multivariate analysis. Br J Cancer 83(1):35–39. https://doi.org/10.1054/bjoc.2000.1199
123. Bonavina L, Soligo D, Quirici N, Bossolasco P, Cesana B, Lembertenghi Deliliers G, Peracchia A (2001) Bone marrow-disseminated tumor cells in patients with carcinoma of the esophagus or cardia. Surgery 129(1):15–22
124. Spence GM, Graham AN, Mulholland K, Maxwell P, McCluggage WG, Sloan JM, McGuigan JA (2004) Bone marrow micrometastases and markers of angiogenesis in esophageal cancer. Ann Thorac Surg 78(6):1944–1949; discussion 1950. https://doi.org/10.1016/j.athoracsur.2004.03.099
125. Ryan P, McCarthy S, Kelly J, Collins JK, Dunne C, Grogan L, Breathnach O, Shanahan F, Carey PD, Walsh TN, O'Sullivan GC (2004) Prevalence of bone marrow micrometastases in esophagogastric cancer patients with and without neoadjuvant chemoradiotherapy. J Surg Res 117(1):121–126. https://doi.org/10.1016/j.jss.2003.12.008
126. Kaifi JT, Yekebas EF, Schurr P, Obonyo D, Wachowiak R, Busch P, Heinecke A, Pantel K, Izbicki JR (2005) Tumor-cell homing to lymph nodes and bone marrow and CXCR4 expression in esophageal cancer. J Natl Cancer Inst 97(24):1840–1847. https://doi.org/10.1093/jnci/dji431
127. Bagheri R, Maddah G, Saedi HS, Sadeghian MH, Roodbari S (2011) Bone marrow involvement in esophageal cancer patients who underwent surgical resection. Eur J Cardiothorac Surg 40(2):343–346. https://doi.org/10.1016/j.ejcts.2010.12.029
128. Gray RT, O'Donnell ME, Verghis RM, McCluggage WG, Maxwell P, McGuigan JA, Spence GM (2012) Bone marrow micrometastases in esophageal carcinoma: a 10-year follow-up study. Dis Esophagus 25(8):709–715. https://doi.org/10.1111/j.1442-2050.2011.01307.x
129. Vashist YK, Effenberger KE, Vettorazzi E, Riethdorf S, Yekebas EF, Izbicki JR, Pantel K (2012) Disseminated tumor cells in bone marrow and the natural course of resected esophageal cancer. Ann Surg 255(6):1105–1112. https://doi.org/10.1097/SLA.0b013e3182565b0b
130. Ryan P, Furlong H, Murphy CG, O'Sullivan F, Walsh TN, Shanahan F, O'Sullivan GC (2015) Prognostic significance of prospectively detected bone marrow micrometastases in esophagogastric cancer: 10-year follow-up confirms prognostic significance. Cancer Med 4(8):1281–1288. https://doi.org/10.1002/cam4.470
131. Wollenberg B, Ollesch A, Maag K, Funke I, Wilmes E (1994) Micrometastases in bone mar-

row of patients with cancers in the head and neck area. Laryngorhinootologie 73(2):88–93. https://doi.org/10.1055/s-2007-997086

132. Gath HJ, Heissler E, Hell B, Bier J, Riethmuller G, Pantel K (1995) Immunocytologic detection of isolated tumor cells in bone marrow of patients with squamous cell carcinomas of the head and neck region. Int J Oral Maxillofac Surg 24(5):351–355
133. Gath HJ, Brakenhoff RH (1999) Minimal residual disease in head and neck cancer. Cancer Metastasis Rev 18(1):109–126
134. Partridge M, Phillips E, Francis R, Li SR (1999) Immunomagnetic separation for enrichment and sensitive detection of disseminated tumour cells in patients with head and neck SCC. J Pathol 189(3):368–377. https://doi.org/10.1002/(SICI)1096-9896(199911)189:3<368::AID-PATH441>3.0.CO;2-2
135. Chaubal S, Wollenberg B, Kastenbauer E, Zeidler R (1999) Ep-CAM--a marker for the detection of disseminated tumor cells in patients suffering from SCCHN. Anticancer Res 19(3B):2237–2242
136. Partridge M, Li SR, Pateromichelakis S, Francis R, Phillips E, Huang XH, Tesfa-Selase F, Langdon JD (2000) Detection of minimal residual cancer to investigate why oral tumors recur despite seemingly adequate treatment. Clin Cancer Res 6(7):2718–2725
137. Partridge M, Brakenhoff R, Phillips E, Ali K, Francis R, Hooper R, Lavery K, Brown A, Langdon J (2003) Detection of rare disseminated tumor cells identifies head and neck cancer patients at risk of treatment failure. Clin Cancer Res 9(14):5287–5294
138. Wollenberg B, Walz A, Kolbow K, Pauli C, Chaubal S, Andratschke M (2004) Clinical relevance of circulating tumour cells in the bone marrow of patients with SCCHN. Onkologie 27(4):358–362. https://doi.org/10.1159/000079088
139. Bleijerveld OB, Brakenhoff RH, Schaaij-Visser TB, Damen JM, Altelaar AF, Heck AJ, Leemans CR, Slijper M (2011) Protein signatures associated with tumor cell dissemination in head and neck cancer. J Proteome 74(4):558–566. https://doi.org/10.1016/j.jprot.2011.01.008
140. Graveland AP, Braakhuis BJ, Eerenstein SE, de Bree R, Bloemena E, de Maaker M, van den Brekel MW, Dijk F, Mesker WE, Tanke HJ, Leemans CR, Brakenhoff RH (2012) Molecular diagnosis of minimal residual disease in head and neck cancer patients. Cell Oncol (Dordr) 35(5):367–375. https://doi.org/10.1007/s13402-012-0097-1
141. Grobe A, Blessmann M, Hanken H, Friedrich RE, Schon G, Wikner J, Effenberger KE, Kluwe L, Heiland M, Pantel K, Riethdorf S (2014) Prognostic relevance of circulating tumor cells in blood and disseminated tumor cells in bone marrow of patients with squamous cell carcinoma of the oral cavity. Clin Cancer Res 20(2):425–433. https://doi.org/10.1158/1078-0432.CCR-13-1101
142. James D. Mueller, Hubert J. Stein, Tao Oyang, Shoji Natsugoe, Marcus Feith, Martin Werner, J. Rüdiger Siewert, Frequency and clinical impact of lymph node micrometastasis and tumor cell microinvolvement in patients with adenocarcinoma of the esophagogastric junction. Cancer 89 (9):1874-1882

Detection of Minimal Residual Disease and Its Clinical Applications in Melanoma and Breast Cancer Patients

5

Selena Y. Lin, Javier I. J. Orozco, and Dave S. B. Hoon

Abstract

Melanoma and breast cancer (BC) patients face a high risk of recurrence and disease progression after curative surgery and/or therapeutic treatment. Monitoring for minimal residual disease (MRD) during a disease-free follow-up period would greatly improve patient outcomes through earlier detection of relapse or treatment resistance. However, MRD monitoring in solid tumors such as melanoma and BC are not well established. Here, we discuss the clinical applications of MRD monitoring in melanoma and BC patients and highlight the current approaches for detecting MRD in these solid tumors.

Keywords

Breast cancer · Melanoma · DTCs · MRD · Liquid biopsy · Micrometastases · cfDNA · cfmiR · CTCs · Biomarkers

5.1 Introduction

With the increasing number of effective cancer therapeutics, screening tests, and technological advancements in minimally-invasive surgeries for solid tumor cancer patients, we continually improve patient outcomes and prognosis. However following curative surgery, cancer patients still face a high risk of recurrence, disease progression, and treatment resistance due to relapse and evolution of distant metastases. The ability to monitor cancer patients after curative surgery and/or therapy for disease progression or recurrence is key to providing earlier detection of cancer for immediate intervention.

The current standard-of-care relies predominately on monitoring clinical manifestations and radiological imaging which are not only harmful to perform on a frequent basis, but it also lacks the necessary sensitivity for detecting recurrent tumors and/or clinical micrometastases particularly when they are small (<1 cm or 10^9 cancer cells) [1]. In the case of leukemia, the presence of only one tumor cell in a background of 100,000 normal cells can have a high level of relapse [2]. Thus, being able to monitor what is termed as minimal residual disease (MRD) is critical in the adjuvant therapy and post-surgery setting. MRD encompasses a wide range of terminology found in the literature including: occult disease, occult micrometastasis, dormant micrometastasis, and/

S. Y. Lin · J. I. J. Orozco · D. S. B. Hoon (✉)
Department of Translational Molecular Medicine, John Wayne Cancer Institute, Santa Monica, CA, USA
e-mail: Selena.Lin@providence.org; HoonD@JWCI.ORG

J. A. Aguirre-Ghiso (ed.), *Biological Mechanisms of Minimal Residual Disease and Systemic Cancer*, Advances in Experimental Medicine and Biology 1100,
https://doi.org/10.1007/978-3-319-97746-1_5

or subclinical disease not measurable by standard clinical imaging methods leading to potential cancer recurrence.

MRD monitoring is well established in hematologic cancers, but not in solid tumors such as melanoma [3] and breast cancer (BC) [4]. Performing repeated biopsy's of solid tumors for assessment of MRD is not practical as it is invasive, limited due to intra-patient tumor heterogeneity, and difficult once the cancer has metastasized to distant organs.

Blood-based MRD monitoring would overcome these limitations. Tumor-derived biomarkers used for MRD monitoring can include detection and characterization of circulating tumor cells (CTCs) and cell-free nucleic acids such as cell-free DNA (cfDNA) and microRNA (cfmiR). These approaches allows for the implementation of applications to monitor early and advanced-stage cancers (Fig. 5.1). Recently other body fluids have been explored such as pleural fluid [5] in lung cancers and urine in genitourinary cancers [6] for potential sources of cancer biomarkers.

Blood-based MRD monitoring for melanoma and BC has several critical clinical applications such as aiding in monitoring for therapeutic efficacy and/or relapse. For example, monitoring solid tumors for MRD on a molecular level can inform clinicians of the intratumor heterogeneity. This would reveal the tumor evolution and/or drug resistance present during longitudinal follow-up allowing for the implementation of better therapeutic regimens to be put into place (i.e. precision medicine). This is increasingly vital as multiple types of targeted therapies and immunotherapies have been approved recently by the FDA for melanoma, and often involve a combination of different treatment regimens not just a single agent. Thus, there is a need to effectively evaluate therapeutic efficacy via MRD assessment. This would greatly improve patient outcomes and enable curative intervention during the earliest possible stages.

The current gold standard for determining prognosis of melanoma upon primary diagnosis involves the American Joint Committee on Cancer (AJCC) staging system and utilizes the sentinel lymph node (SLN) biopsy procedure that is used to assess the degree of metastasis/nodal involvement [7–9]. This standard is limited because there is still variability in the prognosis of melanoma attributed to the heterogeneity of the disease progression [10, 11]. This poses an urgent clinical problem given that once the disease has metastasized to distant organs, patients experience high recurrence rates (<90%). This necessitates the critical need for frequent monitoring of patients.

Blood-based MRD monitoring to enable earlier detection of relapse or metastasis can improve the way patients are followed after curative

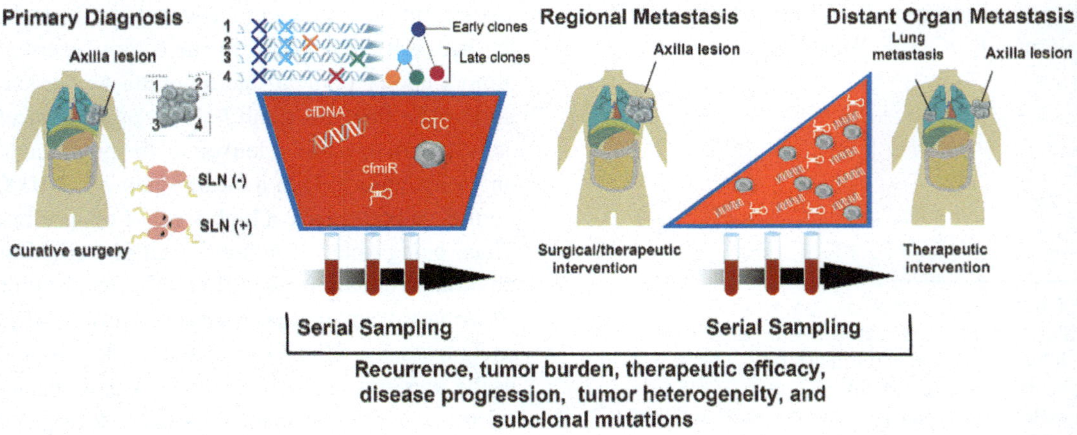

Fig. 5.1 Potential clinical applications for blood-MRD monitoring throughout melanoma disease progression from early to advanced stages. Schematic of serial blood sampling following curative surgery from primary diagnosis of melanoma and following surgical or therapeutic intervention upon relapse or regional metastasis

resection. Unfortunately, there are no existing blood biomarkers for early-stage melanoma to monitor for recurrence. The only AJCC-approved melanoma blood biomarker is serum lactate dehydrogenase (LDH) and is used for prognostication in stage IV melanomas and has been associated with response, progression-free survival (PFS), melanoma-specific survival (MSS), and overall survival (OS) in the contemporary treatment era of targeted and immune therapies.

Monitoring of blood-based MRD in melanoma patients would be ideal for frequent assessment and enables aggressive management of early-stage disease recurrence prior to distant organ metastasis to improve OS. However, the current MRD detection approaches in melanoma and BC still require improvement and evaluation to increase their sensitivity and to better their understand utility. In this chapter, we discuss the current applications and approaches for detecting MRD in melanoma and BCs focusing primarily on CTC and cfNA-derived biomarkers.

5.2 Applications of MRD in Cutaneous Melanoma

One of the most important procedures performed for staging cutaneous melanoma and subsequent identification of the degree of metastases at the time of primary melanoma diagnosis is sentinel lymph node (SLN) biopsy. Much work has been done evaluating the utility of SLN biopsies with numerous clinical trials performed by our group [8, 9, 12–15] and others [16–18]. To date, utilization of SLN biopsies are the current standard for staging cutaneous melanoma and the most important prognostic factor for tumor recurrence and survival in melanoma patients [19, 20].

The objective of SLN biopsies in melanoma are that detection of any number of melanoma cells in one or more lymph nodes that are the first along the lymphatic drainage (i.e. SLN) from the primary cancer site is associated with poorer outcomes. Patients identified with a single metastatic melanoma cell in the SLN potentially develop metastatic disease [21, 22]. Thus, this procedure is informative by identifying the degree of cancer dissemination and allows subsequent upstaging of biopsy-positive patients so they can benefit from further treatment.

SLN biopsies are invasive and examined by histopathological methods, the current gold standard and the assessment terminology of SLN biopsies in regards to classifying microscopic melanoma disease/metastases currently does not account for isolated tumor cells or small cluster of cells (<2 mm) although studies have implicated its importance in melanoma prognosis [23, 24]. Isolated tumor cells (ITC) can be detected with immunohistochemistry (IHC) and by conventional stains such as hematoxylin and eosin (H&E). However, they do not display characteristics of metastases such as proliferation or stromal reaction. It should be noted that MRD terminology is also inconsistent among different cancers. Classifications of SLN biopsies can fall under lymph node (LN) macro- and micrometastases where detection of isolated tumor cells or small cluster of cells (<2 mm) would not be included [10].

Lastly, ITC detected in SLN biopsies may consist of dormant/latent tumor cells that persist during post-remission/disease-free period in melanoma patients. Upon reactivation/re-awakening of dormant melanoma cells after a period ranging anywhere from a year to decades, they can be responsible for melanoma dissemination and distant organ metastasis irrespective of the size of the primary lesion. In addition, dormant cells are resistant to chemotherapies that predominantly target proliferating cells. Adding to heterogeneity of the disease progression, a subset of melanoma patients (~13%) [25, 26] with negative SLN biopsies developed distant organ metastasis and or recurrence without any clinical evidence of regional node positivity at the time of diagnosis. Failure to monitor melanoma patients following initial primary diagnosis and treatment can result in the development of metastatic melanoma, one of the most aggressive cancers. Thus, it is critical during follow-up to monitor for these residual melanoma cells in a more sensitive manner.

5.2.1 Circulating Tumor Cells (CTC)

Blood-based molecular analysis would enable the study and identification of MRD in a minimally-invasive manner. Detection of clinically actionable biomarkers and melanoma dormancy during disease-free follow-up would enable earlier intervention for curative treatment. Furthermore, blood-based monitoring would permit frequent follow-up to assess therapeutic efficacy, disease progression, and potentially capture reactivation of dormant melanoma cells. The most frequently studied melanoma blood-based biomarker assays involve CTCs and cfNA. In general, melanoma CTCs are thought to be responsible for metastatic relapse and disease progression, so detection of CTCs in the blood of melanoma patients would reveal MRD.

CTCs contain intact genomes that are representative of the tumor. This avoids the limitation of cfDNA where genetic material is highly fragmented (<150 bp). Detection of melanoma CTCs encompass either an indirect approach or an antibody-agnostic approach. Indirect analysis of melanoma CTCs utilizes quantitative reverse transcriptase (qRT-PCR) of isolated cells from the peripheral blood without enrichment. CTC analysis for known melanoma mRNA biomarkers (involved in metastases) such as *TRP2, MART1, B4GALNT1, MAGEA3*, and *PAX3* have demonstrated prognostic utility in PFS and OS [7, 27–29] in multicenter phase II/III studies. Typically an mRNA multimarker panel is employed to increase sensitivity of the assay.

An antibody-agnostic approach such as using microfluidics-based CTC enrichment can overcome the hurdles of high CTC heterogeneity and lack of melanoma-specific antibodies. Dual mRNA/DNA biomarker profiling of microfluidics-isolated melanoma CTCs by our group has identified MRD in patients receiving checkpoint inhibitor immunotherapy [30] and highlights the value of comprehensive biomarker profiling to increase not only sensitivity of CTC detection, but also to improve the value of CTC genomic and transcriptomic profiles.

Other approaches for CTC evaluation such as enumeration and phenotypic characterization have also been shown to be clinically useful in providing prognostic information. Melanoma CTC enumeration (≥2/7.5 ml of blood) and monitoring changes of CTC counts during therapy have been reported as prognostic biomarkers correlating with OS [31, 32]. However, these types of assessments are not as informative in a therapeutic setting where identification of dynamic tumor mutation profiles, clonal evolution, and treatment resistance are desired. Furthermore, methodologies have lacked standardization of CTC enumeration analysis, with various cut-offs reported contributing to inconsistent results across different studies. The critical problem of CTC enumeration studies is that there is an assumption that all CTCs are the same with no heterogeneity. This is a major flaw as one cannot assume all CTCs are equal and can establish metastasis.

One of the main difficulties plaguing melanoma CTC enumeration studies is lack of robust technological approaches to isolate rare melanoma CTCs from blood in a high background of normal cells. This makes it particularly difficult to study and evaluate CTCs in early-stage melanoma where detection of residual melanoma disease prior to the on-set of metastasis to distant organs is critical. By contrast, molecular profiling of CTCs to monitor for MRD in patients during follow-up can aid in earlier detection of recurrence and disease progression, and inform therapeutic regimens.

5.2.2 Cell-Free DNA (cfDNA)

Compared to CTCs, cfDNA containing tumor biomarkers is easier to isolate from blood. cfDNA is thought to contain single or double-stranded fragmented (~150 bp) DNA derived from populations of normal and tumor cells. DNA mutations and epigenetic modifications such as DNA methylation derived from cfDNA have been frequently studied in melanoma and its application in identifying MRD can complement detection and enumeration of melanoma CTCs. cfDNA detection of single biomarkers such as *BRAF* or *NRAS* hotspot mutations have been useful in predicting immunotherapy response [33]. Levels of cfDNA mutations have been found to be associated with tumor burden and OS [34–37]. Epigenetic modifications such as Ras association domain family protein 1 (RASSF1A) promoter [38, 39] and estrogen receptor α (ESR α) [40] hypermethylation have also been found to be associated with clinical outcomes.

Like the current CTC assays, cfDNA assays also face the same issue with need for increased sensitivity for detecting cfDNA biomarkers in early-stage melanoma patients where there is high tumor heterogeneity. One approach to overcome tumor heterogeneity is to utilize multi-gene assays with high sensitivity (i.e. 0.1% limit of detection) to detect a panel of cancer-related genes. This has been shown to be useful in earlier detection of recurrence compared to standard radiological imaging where subclones or dynamic tumor populations were captured utilizing a 54-gene digital next generation sequencing (NGS) panel [41]. Another approach is to implement frequent monitoring or serial bleed analysis of cfDNA biomarkers. Because cfDNA has been shown to shed from apoptotic cells, the dynamic levels and assessment of tumor burden can potentially be captured during longitudinal monitoring.

To increase the utility of blood-based cfDNA for monitoring MRD, two factors of assay sensitivity and implementation of serial bleed analysis will need to be improved. Increasing sensitivity of cfDNA assays will continually be addressed with emerging technologies and is critical for early-stage melanomas given that 90% of metastases occur during the first 5-years after diagnosis. In addition, melanoma can have late metastasis so that life-long surveillance is generally recommended particularly if original diagnosis is in the early decades of life [42]. Overall, cfDNA can be clinically useful for earlier detection of melanoma disease prior to distant organ metastasis where for assessing MRD is key.

5.2.3 Cell-Free MicroRNA (cfmiR)

Although less studied in the blood, detection of cfmiRs in blood can also provide valuable information for MRD monitoring. MicroRNAs (miRs) are small non-coding RNAs that have been shown to regulate gene transcription processes such as cell proliferation, differentiation, apoptosis and survival. MicroRNAs also can mediate epigenetic regulation of key genes involved in melanoma metastasis. While it is unclear how detection of cfmiRs will affect therapeutic decision making, it can provide insight to melanoma progression and disease dynamics. One advantage of detecting cfmiRs is that they are extremely stable as they are resistant to endogenous RNAse activity in the blood compared to cfDNA. This decreases the logistical constraints in specimen collection and need for immediate processing and enables the use of miR assays that are not highly sensitive. In fact, we have previously shown through a direct plasma assay (i.e. no extraction required), that cfmiR-210 was a biomarker for disease-free recurrence and was associated with poor outcomes [43]. Detection of elevated levels of cfmiR-221 was shown to be associated with progressive disease compared to melanoma *in situ* and correlated with tumor thickness [44].

Others have used utilized a panel of cfmiRs to demonstrate clinical utility in predicting recurrence risk through a screening/discovery approach [45]. One variable to take into consideration when assessing cfmiRs is its origin given that cfmiRs can be shed from the tumor, or tumor microenvironment, and/or CTCs. Irrespective of the cfmiR source, they all represent a source of biomarkers that can reveal biological and clinical insights into the disease state. Overall, assessing cfmiR as surrogates to monitor MRD can hold prognostic utility and the dynamics of cfmiR through longitudinal monitoring can reveal insight into melanoma disease progression when the disease is present. This can complement the detection of CTCs and cfDNA biomarkers as they each play a unique role in revealing tumor profiles.

5.3 Monitoring of MRD in Breast Cancer

Approximately 20–30% of early-stage breast cancer (EBC) patients develop distant recurrence(s) depending on tumor size, lymph node status upon diagnosis, molecular tumor profile, and therapeutic regimen. The pattern of recurrence (time and site of metastasis) is strongly influenced by the BC molecular subtypes. For example, HER2(+) and triple negative breast cancer (TNBC) exhibit higher rates of visceral and brain metastasis, while bone metastases are more predominant in estrogen receptor (ER)(+) BCs [46]. Furthermore, women with ER(−) BCs have a higher recurrence risk within the first 5 years of diagnosis [47, 48], whereas ER(+) BC patients present a steady rate of distant recurrence ranging from 10% to 40% from 5-years all the way up to 20 years post-initial BC diagnosis [49].

The heterogeneity of BC disease progression attributed to the different BC molecular subtypes exemplifies the need for minimally-invasive and sensitive strategies to monitor BC patients for long-term care. Better informative MRD biomarkers are needed to reduce late BC recurrences and prevent metastasis as metastatic BC remains an incurable disease and a leading cause of cancer-related death in women.

MRD can be detected mainly in regional lymph nodes, in the circulation, and in the bone of BC patients. Axillary lymph node status is a major histological predictor of recurrence risk and is usually evaluated through the SLN biopsy, which minimizes the morbidity of axillary lymph node dissection (ALND) and enables an accurate assessment of MRD without compromising accurate staging or survival [50].

SLN biopsy has allowed for a more detailed pathological evaluation for the detection of MRD, like ITC (malignant cell clusters no larger than 0.2 mm or fewer than 200 cells) or micrometastases (tumor deposits larger than 0.2 mm and/or more than 200 cells, but not greater than 2.0 mm). Since the initial discovery of micrometastases and occult metastases in the SLN, BC studies have shown conflicting results regarding their impact on survival. This further highlights the complexity and heterogeneity of BC disease progression. The NSABP-B32 trial showed that BC patients with occult metastases in SLN, defined as undetectable by standard H&E, but detectable by IHC, in comparison with the absence of occult metastases had significantly worse DFS and OS, but with a very small absolute difference (2.8% for DFS and 1.2% for OS) [51]. In the ACOSOG Z0010 trial, no significant difference was seen in 5-year DFS and OS for those BC patients with occult metastasis in SLN [52]. In a prospective study at our institution, BC patients with micrometastasis in SLN had no differences in 8-year disease-free survival (DFS) or OS compared with SLN-negative patients [53]. This is in contrast with a large population-based study from the Netherlands, where patients with micrometastasis in SLN had a reduced 5-year DFS [54]. In addition, the role of complete ALND of BC patients with SLN micrometastasis compared to SLN-negative patients was found to have no significant impact on DFS and OS in the IBCSG-23-01 trial [55]. Interestingly, some BC patients (27.3%) developed macrometastases in non-SLN as identified in the ACOSOG Z0011 trial [56].

Overall, these conflicting reports of SLN biopsy results with BC patient outcomes point to a larger clinical problem, the need to identify those BC patients who will relapse or go on to develop metastasis. The heterogeneity of disease progression and pathogenesis necessitates sensitive and frequent MRD monitoring to continually assess BC disease progression beyond the initial SLN biopsy procedure.

5.3.1 Disseminated Tumor Cells (DTC) and Circulating Tumor Cells (CTC)

BC metastases are believed to originate from disseminated tumor cells (DTC) that have undergone a period of dormancy [57]. However, the molecular factors that promote the formation of detectable metastases from DTC are largely unknown. At the time of initial diagnosis, DTC can be detected in the bone marrow in 30% of operable BC patients that lack any clinical or histopathological signs of metastasis. A pooled analysis of 4703 patients showed poor outcomes, with shorter DFS and OS in patients with non-metastatic BC with DTC at the time of diagnosis [58]. However, determination of DTC in bone marrow requires bone marrow aspiration or biopsy, a highly invasive procedure. Interestingly, the concordance rate between the synchronous finding of bone-marrow DTC and CTCs in BC has ranged from 66% to 94% in several studies [59], thus highlighting the potential utility of CTCS for evaluating MRD in BC patients.

CTC enumeration using CellSearch®, the only US Food and Drug Administration (FDA)-approved platform for CTC isolation from blood samples of BC patients has been shown to have prognostic value in early and metastatic BC. A baseline of ≥5 CTCs per 7.5 ml of blood, and the persistence of CTCs during treatment was associated with decreased PFS and OS [60, 61]. In a prospective study, the presence of ≥1 CTC (in 24% of patients) predicted early relapse and decreased OS in patients with stage I-III BCs, determined by a single blood collection at the time of definitive surgery [62]. Interestingly, the German phase III SUCCESS trial evaluated CTCs before adjuvant chemotherapy and after chemotherapy in patients with non-metastatic BC. In EBC, adjuvant systemic therapies aim to "empirically" eradicate the presence of micrometastases after surgical resection, diminishing the risk of locoregional recurrences and distant metastasis. The presence of ≥1 CTC before adjuvant chemotherapy could predict poor DFS and OS in patients with EBC [63]. The prognosis was worse in patients with a higher count (≥5 CTC/30 ml). After chemotherapy, CTC were found in 22.1% of patients and were associated with a worse DFS and a trend toward a reduced OS [63]. The identification of MRD after treatment of EBC may identify those patients requiring further treatment or more aggressive disease management. Notably, in a pooled analysis of 3173 patients with non-metastatic BC, the presence of ≥1 CTC at the time of diagnosis was also associated with reduced DFS and OS [64]. However, one should be cautious of considering only a few CTCs as prognostic biomarkers. Accuracy and consistency in pathology-defined true CTC detection is questionable. In parallel with SLN, the presence of a few tumor cells is not considered positive as defined by the pathologist as true breast tumor cells.

Molecular characterization of CTCs in BC also has utility in MRD monitoring. EpCAM-independent quantitative reverse transcriptase PCR (qRT-PCR) for detecting CK-19 mRNA (+) cells was found to be a prognostic factor in several studies [65–69]. Molecular profiling of CTCs has also been shown to reveal insight into treatment efficacy whether it is detecting HER2-status of CTCs [70, 71] or detecting actionable mutations such as in the *PI3KCA* gene [72]. Altogether, this highlights the clinical value of CTCs in all stages of breast cancer for MRD monitoring.

5.3.2 Cell-Free DNA (cfDNA)

Potential cfDNA applications in the context of MRD monitoring are not limited to determining prognosis, monitoring/predicting the response to systemic therapies, and detecting progressive disease before overt clinical and radiological manifestation [73]. Blood cfDNA genomic aberrations that have been previously reported in BC patients include HER2 amplification [74], *TP53* and *PIK3CA* mutations [75], DNA methylation [76], serum DNA integrity [77], and microsatellite alterations [78]. However, the potential to detect tumor-derived cfDNA genomic aberrations during the early-stage BC is challenging, as described in melanoma (Sect. 5. 2.2) due to the low tumor burden of micrometastatic disease. Nonetheless, cfDNA mutations has been evaluated in EBC patients.

The feasibility and cfDNA concordance with surgical tissue specimens have been tested by evaluating *PI3KCA* mutations [79, 80]. The identification of *PI3KCA* mutations in presurgery blood samples had a 93.3% sensitivity and 100% specificity, confirming the utility of cfDNA analysis in patients with EBC [79]. Importantly, cfDNA aberrations identified through serial evaluations could predict early recurrences in patients with non-metastatic BC 7.9–11 months before the diagnosis of metastasis [81, 82]. Therefore, cfDNA detection can be utilized to detect MRD and identify patients with EBC at risk of early relapse. Additionally, these genetic events could be used to define targeted-therapeutic approaches.

5.4 Current Molecular Approaches to Detect MRD

Three molecular approaches frequently used in hematologic malignancies can also be used to detect MRD in solid tumors: real-time PCR (RT-PCR), droplet digital PCR (ddPCR), and NGS. They each hold unique advantages and disadvantages. Many FDA-approved clinical tests utilize RT-PCR detection of single biomarkers because they are simple to set-up and provide rapid diagnostic turn-around times. However, for MRD monitoring where multiplexing for a panel of mRNA or DNA biomarkers maybe needed using the RT-PCR format can be difficult. The next iteration is the ddPCR platform. This offers a more sensitive approach to detect cfNA biomarkers by >100-fold without the need for complicated assay development that may include designing wild-type (WT) suppression oligos (i.e. locked nucleic acid WT clamps). Furthermore, the RainDance Technologies ddPCR platform has been demonstrated to multiplex up to four different KRAS mutations in a single-well [83]. This can greatly improve the sensitivity by reducing the amount of starting input material needed, particularly critical when evaluating DNA methylation biomarkers. One drawback of utilizing RT-PCR and/or ddPCR for monitoring MRD is designing patient-specific assays for assessing tumor subclones. The development of personalized assays serve as a powerful tool in a research setting, but the feasibility in the clinic will need to be addressed. NGS approaches are particularly useful to target a panel of biomarkers that can include cancer hotspot mutations and/or driver genes. DNA mutations and CNVs can be simultaneously assessed and this can improve the sensitivity of detecting early-stage disease where there is high heterogeneity. Digital NGS has further improved the detection sensitivity of these biomarkers for as low as 0.1% variant detection frequency.

Clinical implementation of these approaches will depend on feasibility, assay performance (i.e. sensitivity and specificity), and utility. Certain NGS approaches such as targeted-hybridization maybe lengthy and require 2–3 days for sample processing along with the downstream bioinformatics analysis. Generally MRD monitoring should involve frequent blood sampling and possibly coincide with radiological imaging, although there is no consensus on timing [84]. Blood assays also need to better account for robustness. For example, low tumor-blood cfDNA mutation concordance has plagued cfDNA studies highlighting the lack of "sensitivity" of cfDNA assays. Low concordance can be derived from a biological phenomenon (lack of tumor DNA shedding or rapid cfDNA degradation in blood), assay robustness, or sampling inefficiency. This urges the need to perform frequent blood sampling in an attempt to overcome these hurdles. Exacerbated by the technical hurdles described above, single blood analysis or pre and post-surgery analysis of blood biomarkers for assessing MRD are now known to be insufficient.

Lastly, approaches used to evaluate CTCs isolated from blood can include enumeration or molecular characterization approaches described above. As the CTC field has evolved the caveat of utilization of CTC enumeration is limited clinical value. Techniques that enumerate isolated CTCs have shown conflicting and non-reproducible results. Furthermore, CTCs are heterogeneous as seen in melanoma, further limiting the value of enumeration in clinical decision making. Overall, the current approaches to assess MRD have focused on sensitive molecular characterization with clinically actionable biomarkers.

5.5 Future Outlook

MRD monitoring can enable tremendous improvement in the way melanoma and BC patients are monitored following surgical resection or therapeutic treatment. Earlier detection of regional and distant metastases will improve patient outcomes. However moving forward, several hurdles will need to be overcomed. Investigation into the timing of imaging along with MRD assessments during follow-up will need to be done to demonstrate the optimal balance between economic costs and clinical utility. Given that multiple time points and serial sampling are needed to detect MRD in the blood, the costs of particular assays will need to be considered. Furthermore, when MRD is detected, but patients are deemed clinically disease-free by standard imaging approaches the impact on clinical decision-making will need to be determined.

In addition, biomarker sensitivity, specificity, and clinical utility will need to be robustly validated. Detection of drug-able targets in blood is relatively straightforward because they may be used to guide treatment regimens and enable patient treatment stratification. Detection of other targets such as detection of *TP53* mutations or up-regulated cfmiRs require more work and need prospective patient cohorts to demonstrate utility (i.e. whether they can serve as diagnostic or prognostic biomarkers). The problem is that in a single or few cfmiR detection assays the specificity and sensitivity is lacking when using standard PCR assay approaches. The issue of false positive and negative results will also need to be addressed as more sensitive molecular approaches are developed and how these results will ultimately affect clinical decision-making. Lastly, as the biomarkers identified in CTCs, cfDNA, and cfmiR are further investigated, new emerging biomarkers should also be considered such as exosomes. They may be a valuable source of MRD biomarkers as they hold critical biological molecules for intercellular crosstalk and are responsible for the transfer of functional molecular cargo to recipient cells.

5.6 Summary

Although blood-based biomarkers are not well established in the context of MRD monitoring for melanoma and breast cancer, its clinical utility is vital to improving patient outcomes for early and/or late stages of disease. MRD monitoring using blood-based biomarkers, performed in a frequent and repetitive manner will enable assessment of the tumor burden, progression, dynamic molecular profile, and detection of subclones. This will enable clinicians to detect relapse and/or distant organ metastasis earlier than standard radiological imaging. This ultimately will provide earlier intervention to achieve curative treatment at the earliest possible stages. Overall, more work is needed to robustly demonstrate the utility of blood-based MRD monitoring in melanoma and BC patients, but this approach holds great promise for early detection of the disease.

Acknowledgements We would like to acknowledge Dr. Ian Hutchinson, Editorial Services, for his support in editing of this manuscript, and Rebecca Gentry for her assistance in proofreading the manuscript.

References

1. Frangioni JV (2008) New technologies for human cancer imaging. J Clin Oncol 26:4012–4021
2. Stow P, Key L, Chen X, Pan Q, Neale GA, Coustan-Smith E, Mullighan CG, Zhou Y, Pui C-H, Campana D (2010) Clinical significance of low levels of minimal residual disease at the end of remission induction therapy in childhood acute lymphoblastic leukemia. Blood 115:4657–4663
3. Trotter SC, Sroa N, Winkelmann RR, Olencki T, Bechtel M (2013) A global review of melanoma follow-up guidelines. J Clin Aesthet dermatol 6:18
4. Khatcheressian JL, Hurley P, Bantug E, Esserman LJ, Grunfeld E, Halberg F, Hantel A, Henry NL, Muss HB, Smith TJ (2012) Breast cancer follow-up and management after primary treatment: American Society of Clinical Oncology clinical practice guideline update. J Clin Oncol 31:961–965
5. Kawahara A, Fukumitsu C, Azuma K, Taira T, Abe H, Takase Y, Murata K, Sadashima E, Hattori S, Naito Y (2018) A combined test using both cell sediment and supernatant cell-free DNA in pleural effusion shows increased sensitivity in detecting activating EGFR mutation in lung cancer patients. Cytopathology 29(2):150–155
6. Lin SY, Linehan JA, Wilson TG, Hoon DS (2017) Emerging utility of urinary cell-free nucleic acid biomarkers for prostate, bladder, and renal cancers. Eur Urol Focus 3:265–272
7. Hoshimoto S, Shingai T, Morton DL, Kuo C, Faries MB, Chong K, Elashoff D, Wang H-J, Elashoff RM,

Hoon DS (2012) Association between circulating tumor cells and prognosis in patients with stage III melanoma with sentinel lymph node metastasis in a phase III international multicenter trial. J Clin Oncol 30:3819–3826
8. Morton DL, Thompson JF, Cochran AJ, Mozzillo N, Elashoff R, Essner R, Nieweg OE, Roses DF, Hoekstra HJ, Karakousis CP (2006) Sentinel-node biopsy or nodal observation in melanoma. N Engl J Med 355:1307–1317
9. Morton DL, Thompson JF, Cochran AJ, Mozzillo N, Nieweg OE, Roses DF, Hoekstra HJ, Karakousis CP, Puleo CA, Coventry BJ (2014) Final trial report of sentinel-node biopsy versus nodal observation in melanoma. N Engl J Med 370:599–609
10. Balch CM, Gershenwald JE, Soong S-J, Thompson JF, Atkins MB, Byrd DR, Buzaid AC, Cochran AJ, Coit DG, Ding S (2009) Final version of 2009 AJCC melanoma staging and classification. J Clin Oncol 27:6199–6206
11. Balch CM, Gershenwald JE, Soong S-J, Thompson JF, Ding S, Byrd DR, Cascinelli N, Cochran AJ, Coit DG, Eggermont AM (2010) Multivariate analysis of prognostic factors among 2,313 patients with stage III melanoma: comparison of nodal micrometastases versus macrometastases. J Clin Oncol 28:2452–2459
12. Faries MB, Thompson JF, Cochran AJ, Andtbacka RH, Mozzillo N, Zager JS, Jahkola T, Bowles TL, Testori A, Beitsch PD, Hoekstra HJ, Moncrieff M, Ingvar C, Wouters MWJM, Sabel MS, Levine EA, Agnese D, Henderson M, Dummer R, Rossi CR, Neves RI, Trocha SD, Wright F, Byrd DR, Matter M, Hsueh E, Mackenzie-Ross A, Johnson DB, Terheyden P, Berger AC, Huston TL, Wayne JD, Smithers BM, Neuman HB, Schneebaum S, Gershenwald JE, Ariyan CE, Desai DC, Jacobs L, Mcmasters KM, Gesierich A, Hersey P, Bines SD, Kane JM, Barth RJ, Mckinnon G, Farma JM, Schultz E, Vidal-Sicart S, Hoefer RA, Lewis JM, Scheri R, Kelley MC, Nieweg OE, Noyes RD, Hoon DSB, Wang H-J, Elashoff DA, Elashoff RM (2017) Completion dissection or observation for sentinel-node metastasis in melanoma. N Engl J Med 376:2211–2222
13. Morton DL (2012) Overview and update of the phase III multicenter selective lymphadenectomy trials (MSLT-I and MSLT-II) in melanoma. Clin Exp Metastasis 29:699–706
14. Nicholl MB, Elashoff D, Takeuchi H, Morton DL, Hoon DS (2011) Molecular upstaging based on paraffin-embedded sentinel lymph nodes: ten-year follow-up confirms prognostic utility in melanoma patients. Ann Surg 253:116
15. Wong SL, Morton DL, Thompson JF, Gershenwald JE, Leong SP, Reintgen DS, Gutman H, Sabel MS, Carlson GW, Mcmasters KM (2006) Melanoma patients with positive sentinel nodes who did not undergo completion lymphadenectomy: a multi-institutional study. Ann Surg Oncol 13:809–816
16. Leiter U, Stadler R, Mauch C, Hohenberger W, Brockmeyer N, Berking C, Sunderkötter C, Kaatz M, Schulte K-W, Lehmann P (2015) Survival of SLNB-positive melanoma patients with and without complete lymph node dissection: A multicenter, randomized DECOG trial. Am Soc Clin Oncol 17:757–767
17. Leiter U, Stadler R, Mauch C, Hohenberger W, Brockmeyer N, Berking C, Sunderkötter C, Kaatz M, Schulte K-W, Lehmann P (2016) Complete lymph node dissection versus no dissection in patients with sentinel lymph node biopsy positive melanoma (DeCOG-SLT): a multicentre, randomised, phase 3 trial. Lancet Oncol 17:757–767
18. Melstrom LG, Taylor E, Kuk D, Frankel TL, Panageas K, Haydu L, Sabel MS, Thompson JF, Ariyan C, Coit DG (2014) International multi-institutional management and outcome of melanoma patients with positive sentinel lymph nodes in more than one nodal basin. Ann Surg Oncol 21:4324–4329
19. Balch CM, Morton DL, Gershenwald JE, Mcmasters KM, Nieweg OE, Powell B, Ross MI, Sondak VK, Thompson JF (2009) Sentinel node biopsy and standard of care for melanoma. J Am Acad Dermatol 60:872–875
20. Gershenwald JE, Scolyer RA, Hess KR, Sondak VK, Long GV, Ross MI, Lazar AJ, Faries MB, Kirkwood JM, McArthur GA (2017) Melanoma staging: evidence-based changes in the American joint committee on Cancer eighth edition cancer staging manual. CA Cancer J Clin 67(6):472–492
21. Murali R, Desilva C, McCarthy SW, Thompson JF, Scolyer RA (2012) Sentinel lymph nodes containing very small (< 0.1 mm) deposits of metastatic melanoma cannot be safely regarded as tumor-negative. Ann Surg Oncol 19:1089–1099
22. Scheri RP, Essner R, Turner RR, Ye X, Morton DL (2007) Isolated tumor cells in the sentinel node affect long-term prognosis of patients with melanoma. Ann Surg Oncol 14:2861–2866
23. Ulmer A, Dietz K, Hodak I, Polzer B, Scheitler S, Yildiz M, Czyz Z, Lehnert P, Fehm T, Hafner C (2014) Quantitative measurement of melanoma spread in sentinel lymph nodes and survival. PLoS Med 11:e1001604
24. Werner-Klein M, Scheitler S, Hoffmann M, Hodak I, Dietz K, Lehnert P, Naimer V, Polzer B, Treitschke S, Werno C (2018) Genetic alterations driving metastatic colony formation are acquired outside of the primary tumour in melanoma. Nat Commun 9:595
25. Carlson GW, Page AJ, Cohen C, Parker D, Yaar R, Li A, Hestley A, Delman KA, Murray DR (2008) Regional recurrence after negative sentinel lymph node biopsy for melanoma. Ann Surg 248:378–386
26. Gershenwald JE, Colome MI, Lee JE, Mansfield PF, Tseng C-H, Lee JJ, Balch CM, Ross MI (1998) Patterns of recurrence following a negative sentinel lymph node biopsy in 243 patients with stage I or II melanoma. J Clin Oncol 16:2253–2260
27. Hoshimoto S, Faries MB, Morton DL, Shingai T, Kuo C, Wang H-J, Elashoff R, Mozzillo N, Kelley MC, Thompson JF (2012) Assessment of prognostic circulating tumor cells in a phase III trial of adjuvant

immunotherapy after complete resection of stage IV melanoma. Ann Surg 255:357
28. Koyanagi K, O'Day SJ, Boasberg P, Atkins MB, Wang H-J, Gonzalez R, Lewis K, Thompson JA, Anderson CM, Lutzky J (2010) Serial monitoring of circulating tumor cells predicts outcome of induction biochemotherapy plus maintenance biotherapy for metastatic melanoma. Clin Cancer Res 16:2402–2408
29. Koyanagi K, O'Day SJ, Gonzalez R, Lewis K, Robinson WA, Amatruda TT, Wang H-J, Elashoff RM, Takeuchi H, Umetani N (2005) Serial monitoring of circulating melanoma cells during neoadjuvant biochemotherapy for stage III melanoma: outcome prediction in a multicenter trial. J Clin Oncol 23:8057–8064
30. Lin S, Lam S, Ohe S, Tran K, Ramos I, Bhagat AA, Lim CT, O'Day SJ, Foshag L, Hoon D (2017) Mo'nitoring of multimodality immune checkpoint inhibitor treatment efficacy in metastatic melanoma patients through molecular analysis of circulating tumor cells. AACR
31. Khoja L, Lorigan P, Zhou C, Lancashire M, Booth J, Cummings J, Califano R, Clack G, Hughes A, Dive C (2013) Biomarker utility of circulating tumor cells in metastatic cutaneous melanoma. J Investig Dermatol 133:1582–1590
32. Klinac D, Gray ES, Freeman JB, Reid A, Bowyer S, Millward M, Ziman M (2014) Monitoring changes in circulating tumour cells as a prognostic indicator of overall survival and treatment response in patients with metastatic melanoma. BMC Cancer 14:423
33. Lee J, Long G, Boyd S, Lo S, Menzies A, Tembe V, Guminski A, Jakrot V, Scolyer R, Mann G (2017) Circulating tumour DNA predicts response to anti-PD1 antibodies in metastatic melanoma. Ann Oncol 28:1130–1136
34. Gray ES, Rizos H, Reid AL, Boyd SC, Pereira MR, Lo J, Tembe V, Freeman J, Lee JH, Scolyer RA (2015) Circulating tumor DNA to monitor treatment response and detect acquired resistance in patients with metastatic melanoma. Oncotarget 6:42008
35. Khagi Y, Goodman AM, Daniels GA, Patel SP, Sacco AG, Randall JM, Bazhenova LA, Kurzrock R (2017) Hypermutated circulating tumor DNA: correlation with response to checkpoint inhibitor–based immunotherapy. Clin Cancer Res 23:5729–5736
36. Lee R, Gremel G, Marshall A, Myers KA, Fisher N, Dunn J, Dhomen N, Corrie P, Middleton MR, Lorigan P (2017) Circulating tumor DNA predicts survival in patients with resected high risk stage II/III melanoma. Ann Oncol 29(2):490–496
37. Shinozaki M, O'Day SJ, Kitago M, Amersi F, Kuo C, Kim J, Wang H-J, Hoon DS (2007) Utility of circulating B-RAF DNA mutation in serum for monitoring melanoma patients receiving biochemotherapy. Clin Cancer Res 13:2068–2074
38. Koyanagi K, Mori T, O'Day SJ, Martinez SR, Wang H-J, Hoon DS (2006) Association of circulating tumor cells with serum tumor-related methylated DNA in peripheral blood of melanoma patients. Cancer Res 66:6111–6117
39. Mori T, O'Day S, Umetani N, Kuo C, Kitago M, Wilson L, Martinez S, Takeshima T-L, Milford R, Hoon D (2005) Circulating methylated DNA in serum predicts melanoma patients response to biochemotherapy. J Clin Oncol 23:7521–7521
40. Mori T, Martinez SR, O'Day SJ, Morton DL, Umetani N, Kitago M, Tanemura A, Nguyen SL, Tran AN, Wang H-J (2006) Estrogen receptor-α methylation predicts melanoma progression. Cancer Res 66:6692–6698
41. Lin SY, Huang SK, Huynh Kelly T, Salomon MP, Chang S-C, Marzese D, Lanman R, Talasaz A, Hoon DS (2018) Multiplex Gene Profiling of Cell-Free DNA in metastatic melanoma patients for monitoring disease. J Clin Oncol Precision Oncol. Manuscript Accepted
42. Garbe C, Schadendorf D, Stolz W, Volkenandt M, Reinhold U, Kortmann RD, Kettelhack C, Frerich B, Keilholz U, Dummer R (2008) Short German guidelines: malignant melanoma. JDDG: J Dtch Dermatol Ges 6:S9–S14
43. Ono S, Oyama T, Lam S, Chong K, Foshag LJ, Hoon DS (2015) A direct plasma assay of circulating microRNA-210 of hypoxia can identify early systemic metastasis recurrence in melanoma patients. Oncotarget 6:7053
44. Kanemaru H, Fukushima S, Yamashita J, Honda N, Oyama R, Kakimoto A, Masuguchi S, Ishihara T, Inoue Y, Jinnin M (2011) The circulating microRNA-221 level in patients with malignant melanoma as a new tumor marker. J Dermatol Sci 61:187–193
45. Friedman EB, Shang S, de Miera EV-S, Fog JU, Teilum MW, Ma MW, Berman RS, Shapiro RL, Pavlick AC, Hernando E (2012) Serum microRNAs as biomarkers for recurrence in melanoma. J Transl Med 10:155
46. Kennecke H, Yerushalmi R, Woods R, Cheang MC, Voduc D, Speers CH, Nielsen TO, Gelmon K (2010) Metastatic behavior of breast cancer subtypes. J Clin Oncol 28:3271–3277
47. Metzger-Filho O, Sun Z, Viale G, Price KN, Crivellari D, Snyder RD, Gelber RD, Castiglione-Gertsch M, Coates AS, Goldhirsch A, Cardoso F (2013) Patterns of recurrence and outcome according to breast cancer subtypes in lymph node-negative disease: results from international breast cancer study group trials VIII and IX. J Clin Oncol 31:3083–3090
48. Saphner T, Tormey DC, Gray R (1996) Annual hazard rates of recurrence for breast cancer after primary therapy. J Clin Oncol 14:2738–2746
49. Pan H, Gray R, Braybrooke J, Davies C, Taylor C, Mcgale P, Peto R, Pritchard KI, Bergh J, Dowsett M, Hayes DF (2017) 20-year risks of breast-cancer recurrence after stopping endocrine therapy at 5 years. N Engl J Med 377:1836–1846
50. Mamounas EP, Kuehn T, Rutgers EJT, von Minckwitz G (2017) Current approach of the axilla in patients with early-stage breast cancer. Lancet

51. Weaver DL, Ashikaga T, Krag DN, Skelly JM, Anderson SJ, Harlow SP, Julian TB, Mamounas EP, Wolmark N (2011) Effect of occult metastases on survival in node-negative breast cancer. N Engl J Med 364:412–421
52. Giuliano AE, Hawes D, Ballman KV, Whitworth PW, Blumencranz PW, Reintgen DS, Morrow M, Leitch AM, Hunt KK, McCall LM, Abati A, Cote R (2011) Association of occult metastases in sentinel lymph nodes and bone marrow with survival among women with early-stage invasive breast cancer. JAMA 306:385–393
53. Hansen NM, Grube B, Ye X, Turner RR, Brenner RJ, Sim MS, Giuliano AE (2009) Impact of micrometastases in the sentinel node of patients with invasive breast cancer. J Clin Oncol 27:4679–4684
54. de Boer M, van Deurzen CH, van Dijck JA, Borm GF, van Diest PJ, Adang EM, Nortier JW, Rutgers EJ, Seynaeve C, Menke-Pluymers MB, Bult P, Tjan-Heijnen VC (2009) Micrometastases or isolated tumor cells and the outcome of breast cancer. N Engl J Med 361:653–663
55. Galimberti V, Cole BF, Zurrida S, Viale G, Luini A, Veronesi P, Baratella P, Chifu C, Sargenti M, Intra M, Gentilini O, Mastropasqua MG, Mazzarol G, Massarut S, Garbay JR, Zgajnar J, Galatius H, Recalcati A, Littlejohn D, Bamert M, Colleoni M, Price KN, Regan MM, Goldhirsch A, Coates AS, Gelber RD, Veronesi U (2013) Axillary dissection versus no axillary dissection in patients with sentinel-node micrometastases (IBCSG 23-01): a phase 3 randomised controlled trial. Lancet Oncol 14:297–305
56. Giuliano AE, Ballman KV, McCall L, Beitsch PD, Brennan MB, Kelemen PR, Ollila DW, Hansen NM, Whitworth PW, Blumencranz PW, Leitch AM, Saha S, Hunt KK, Morrow M (2017) Effect of axillary dissection vs no axillary dissection on 10-year overall survival among women with invasive breast cancer and sentinel node metastasis: the ACOSOG Z0011 (alliance) randomized clinical trial. JAMA 318:918–926
57. Aguirre-Ghiso JA (2007) Models, mechanisms and clinical evidence for cancer dormancy. Nat Rev Cancer 7:834–846
58. Braun S, Vogl FD, Naume B, Janni W, Osborne MP, Coombes RC, Schlimok G, Diel IJ, Gerber B, Gebauer G, Pierga JY, Marth C, Oruzio D, Wiedswang G, Solomayer EF, Kundt G, Strobl B, Fehm T, Wong GY, Bliss J, Vincent-Salomon A, Pantel K (2005) A pooled analysis of bone marrow micrometastasis in breast cancer. N Engl J Med 353:793–802
59. Bidard FC, Proudhon C, Pierga JY (2016) Circulating tumor cells in breast cancer. Mol Oncol 10:418–430
60. Bidard FC, Peeters DJ, Fehm T, Nole F, Gisbert-Criado R, Mavroudis D, Grisanti S, Generali D, Garcia-Saenz JA, Stebbing J, Caldas C, Gazzaniga P, Manso L, Zamarchi R, de Lascoiti AF, de Mattos-Arruda L, Ignatiadis M, Lebofsky R, van Laere SJ, Meier-Stiegen F, Sandri MT, Vidal-Martinez J, Politaki E, Consoli F, Bottini A, Diaz-Rubio E, Krell J, Dawson SJ, Raimondi C, Rutten A, Janni W, Munzone E, Caranana V, Agelaki S, Almici C, Dirix L, Solomayer EF, Zorzino L, Johannes H, Reis-Filho JS, Pantel K, Pierga JY, Michiels S (2014) Clinical validity of circulating tumour cells in patients with metastatic breast cancer: a pooled analysis of individual patient data. Lancet Oncol 15:406–414
61. Cristofanilli M, Budd GT, Ellis MJ, Stopeck A, Matera J, Miller MC, Reuben JM, Doyle GV, Allard WJ, Terstappen LW, Hayes DF (2004) Circulating tumor cells, disease progression, and survival in metastatic breast cancer. N Engl J Med 351:781–791
62. Lucci A, Hall CS, Lodhi AK, Bhattacharyya A, Anderson AE, Xiao L, Bedrosian I, Kuerer HM, Krishnamurthy S (2012) Circulating tumour cells in non-metastatic breast cancer: a prospective study. Lancet Oncol 13:688–695
63. Rack B, Schindlbeck C, Juckstock J, Andergassen U, Hepp P, Zwingers T, Friedl TW, Lorenz R, Tesch H, Fasching PA, Fehm T, Schneeweiss A, Lichtenegger W, Beckmann MW, Friese K, Pantel K, Janni W (2014) Circulating tumor cells predict survival in early average-to-high risk breast cancer patients. J Natl Cancer Inst 106
64. Janni WJ, Rack B, Terstappen LW, Pierga JY, Taran FA, Fehm T, Hall C, de Groot MR, Bidard FC, Friedl TW, Fasching PA, Brucker SY, Pantel K, Lucci A (2016) Pooled analysis of the prognostic relevance of circulating tumor cells in primary breast cancer. Clin Cancer Res 22:2583–2593
65. Ignatiadis M, Xenidis N, Perraki M, Apostolaki S, Politaki E, Kafousi M, Stathopoulos EN, Stathopoulou A, Lianidou E, Chlouverakis G (2007) Different prognostic value of cytokeratin-19 mrna–positive circulating tumor cells according to estrogen receptor and HER2 status in early-stage breast cancer. J Clin Oncol 25:5194–5202
66. Saloustros E, Perraki M, Apostolaki S, Kallergi G, Xyrafas A, Kalbakis K, Agelaki S, Kalykaki A, Georgoulias V, Mavroudis D (2011) Cytokeratin-19 mRNA-positive circulating tumor cells during follow-up of patients with operable breast cancer: prognostic relevance for late relapse. Breast Cancer Res 13:R60
67. Xenidis N, Ignatiadis M, Apostolaki S, Perraki M, Kalbakis K, Agelaki S, Stathopoulos EN, Chlouverakis G, Lianidou E, Kakolyris S (2009) Cytokeratin-19 mRNA-positive circulating tumor cells after adjuvant chemotherapy in patients with early breast cancer. J Clin Oncol 27:2177–2184
68. Xenidis N, Markos V, Apostolaki S, Perraki M, Pallis A, Sfakiotaki G, Papadatos-Pastos D, Kalmanti L, Kafousi M, Stathopoulos E (2007) Clinical relevance of circulating CK-19 mRNA-positive cells detected during the adjuvant tamoxifen treatment in patients with early breast cancer. Ann Oncol 18:1623–1631
69. Xenidis N, Perraki M, Kafousi M, Apostolaki S, Bolonaki I, Stathopoulou A, Kalbakis K, Androulakis N, Kouroussis C, Pallis T (2006) Predictive and prognostic value of peripheral blood cytokeratin-19 mRNA-positive cells detected by real-time

polymerase chain reaction in node-negative breast cancer patients. J Clin Oncol 24:3756–3762

70. Agelaki S, Kalykaki A, Markomanolaki H, Papadaki MA, Kallergi G, Hatzidaki D, Kalbakis K, Mavroudis D, Georgoulias V (2015) Efficacy of lapatinib in therapy-resistant HER2-positive circulating tumor cells in metastatic breast cancer. PLoS One 10:e0123683
71. Georgoulias V, Bozionelou V, Agelaki S, Perraki M, Apostolaki S, Kallergi G, Kalbakis K, Xyrafas A, Mavroudis D (2012) Trastuzumab decreases the incidence of clinical relapses in patients with early breast cancer presenting chemotherapy-resistant CK-19mRNA-positive circulating tumor cells: results of a randomized phase II study. Ann Oncol 23:1744–1750
72. Schneck H, Blassl C, Meier-Stiegen F, Neves RP, Janni W, Fehm T, Neubauer H (2013) Analysing the mutational status of PIK3CA in circulating tumor cells from metastatic breast cancer patients. Mol Oncol 7:976–986
73. de Mattos-Arruda L, Caldas C (2016) Cell-free circulating tumour DNA as a liquid biopsy in breast cancer. Mol Oncol 10:464–474
74. Page K, Hava N, Ward B, Brown J, Guttery D, Ruangpratheep C, Blighe K, Sharma A, Walker R, Coombes R (2011) Detection of HER2 amplification in circulating free DNA in patients with breast cancer. Br J Cancer 104:1342
75. Dawson S-J, Tsui DW, Murtaza M, Biggs H, Rueda OM, Chin S-F, Dunning MJ, Gale D, Forshew T, Mahler-Araujo B (2013) Analysis of circulating tumor DNA to monitor metastatic breast cancer. N Engl J Med 368:1199–1209
76. Sharma G, Mirza S, Parshad R, Srivastava A, Gupta SD, Pandya P, Ralhan R (2010) Clinical significance of promoter hypermethylation of DNA repair genes in tumor and serum DNA in invasive ductal breast carcinoma patients. Life Sci 87:83–91
77. Umetani N, Giuliano AE, Hiramatsu SH, Amersi F, Nakagawa T, Martino S, Hoon DS (2006) Prediction of breast tumor progression by integrity of free circulating DNA in serum. J Clin Oncol 24:4270–4276
78. Shaw JA, Page K, Blighe K, Hava N, Guttery D, Ward B, Brown J, Ruangpratheep C, Stebbing J, Payne R (2012) Genomic analysis of circulating cell-free DNA infers breast cancer dormancy. Genome Res 22:220–231
79. Beaver JA, Jelovac D, Balukrishna S, Cochran R, Croessmann S, Zabransky DJ, Wong HY, Toro PV, Cidado J, Blair BG, Chu D, Burns T, Higgins MJ, Stearns V, Jacobs L, Habibi M, Lange J, Hurley PJ, Lauring J, Vandenberg D, Kessler J, Jeter S, Samuels ML, Maar D, Cope L, Cimino-Mathews A, Argani P, Wolff AC, Park BH (2014) Detection of cancer DNA in plasma of patients with early-stage breast cancer. Clin Cancer Res 20:2643–2650
80. Board RE, Wardley AM, Dixon JM, Armstrong AC, Howell S, Renshaw L, Donald E, Greystoke A, Ranson M, Hughes A, Dive C (2010) Detection of PIK3CA mutations in circulating free DNA in patients with breast cancer. Breast Cancer Res Treat 120:461–467
81. Garcia-Murillas I, Schiavon G, Weigelt B, Ng C, Hrebien S, Cutts RJ, Cheang M, Osin P, Nerurkar A, Kozarewa I, Garrido JA, Dowsett M, Reis-Filho JS, Smith IE, Turner NC (2015) Mutation tracking in circulating tumor DNA predicts relapse in early breast cancer. Sci Transl Med 7:302ra133
82. Olsson E, Winter C, George A, Chen Y, Howlin J, Tang MH, Dahlgren M, Schulz R, Grabau D, van Westen D, Ferno M, Ingvar C, Rose C, Bendahl PO, Ryden L, Borg A, Gruvberger-Saal SK, Jernstrom H, Saal LH (2015) Serial monitoring of circulating tumor DNA in patients with primary breast cancer for detection of occult metastatic disease. EMBO Mol Med 7:1034–1047
83. Taly V, Pekin D, Benhaim L, Kotsopoulos SK, Le Corre D, Li X, Atochin I, Link DR, Griffiths AD, Pallier K (2013) Multiplex picodroplet digital PCR to detect KRAS mutations in circulating DNA from the plasma of colorectal cancer patients. Clin Chem 59:1722–1731
84. Grange F, Vitry F, Granel-Brocard F, Lipsker D, Aubin F, Hédelin G, Dalac S, Truchetet F, Michel C, Batard M-L (2008) Variations in management of stage I to stage III cutaneous melanoma: a population-based study of clinical practices in France. Arch Dermatol 144:629–636

Preservation of Quiescent Chronic Myelogenous Leukemia Stem Cells by the Bone Marrow Microenvironment

6

Mansi Shah and Ravi Bhatia

Abstract

The majority of leukemia patients achieving remission ultimately relapse. Persistence of leukemia stem cells (LSC) capable of regenerating leukemia is a major cause of relapse. There is a pressing need to better understand mechanisms of LSC regulation and their resistance to therapy in order to improve outcomes for leukemia. Chronic myelogenous leukemia (CML) is a lethal myeloproliferative disorder that that is caused by hematopoietic stem cell (HSC) transformation by the BCR-ABL tyrosine kinase. Treatment with tyrosine kinase inhibitors (TKI) has revolutionized CML treatment, but fails to eliminate LSC responsible for propagating and regenerating leukemia. Therefore, patients require continued treatment to prevent relapse. Leukemic and normal stem cells share properties of quiescence and self-renewal, that are supported by bone marrow niches. Persistence of LSC after TKI treatment is related to tyrosine kinase independent mechanisms which include intrinsic properties of LSCs determined by epigenetic alterations, altered transcriptional regulatory networks or mitochondrial/metabolic changes. In addition to cell intrinsic changes, signals from the bone marrow microenvironment (BMM) play a critical role in protecting LSC from TKI treatment. Each type of alteration may offer potential points of intervention for therapeutic targeting of LSC.

Keywords

CML · Leukemic stem cell · BCR-ABL · Hematopoietic stem cell · Myeloproliferative · Bone marrow · Microenvironment · Mesenchymal stromal stem cells · Chemokine · Niches

6.1 Introduction

The majority of leukemia patients achieving remission are not cured and ultimately relapse. Relapse is the major contributor to leukemia-related mortality and morbidity, and its human, societal and financial costs. The inability to cure leukemia patients is related to genetic and functional heterogeneity of leukemia cells. Thus, persistence of leukemia stem cells (LSC) capable of regenerating leukemia is a major cause of relapse. LSC share properties with normal HSC and resist elimination by conventional and targeted therapies. There is a pressing need to better understand mechanisms of LSC regulation and their

M. Shah · R. Bhatia (✉)
Division of Hematology & Oncology, Department of Medicine, University of Alabama at Birmingham, Birmingham, AL, USA
e-mail: rbhatia@uabmc.edu

J. A. Aguirre-Ghiso (ed.), *Biological Mechanisms of Minimal Residual Disease and Systemic Cancer*, Advances in Experimental Medicine and Biology 1100,
https://doi.org/10.1007/978-3-319-97746-1_6

resistance to therapy in order to improve outcomes for leukemia.

Chronic myelogenous leukemia (CML) is a lethal myeloproliferative disorder that that is caused by hematopoietic stem cell (HSC) transformation by the BCR-ABL tyrosine kinase [1]. CML progresses from initial chronic phase (CP) to accelerated phase (AP) and terminal blast crisis (BC). Treatment with tyrosine kinase inhibitors (TKI) has revolutionized CML treatment, and results in deep remissions and prolongs survival in most CML patient [2]. However, TKI treatment fails to eliminate LSC responsible for propagating and regenerating leukemia [3]. Therefore, treatment discontinuation usually results in the rapid relapse of leukemia, and most patients require continued treatment to prevent relapse. Several studies have shown that TKIs effectively inhibit kinase activity within LSC, indicating that resistance occurs through “kinase-independent” mechanisms [4, 5]. TKI inhibit proliferation of LSC, but do not efficiently induce apoptosis, with quiescent LSC being especially resistant to TKI treatment [6, 7]. In addition to cell intrinsic changes, signals from the bone marrow microenvironment (BMM) play a critical role in protecting LSC from TKI treatment [8].

Risks of prolonged TKI treatment include non-compliance with risk of relapse and progression; toxicity, including serious vascular complications; and teratogenic effects precluding pregnancy while on treatment [9]. Importantly, the high costs of TKIs has a huge financial impact, compounded by increasing cost of drugs and increasing prevalence of CML in the population. In addition to the societal burden, high copayment requirements impact individual patients and adversely affect adherence to treatment. Persistent LSC in addition to being the reason for lack of cure, may also be the ultimate source of acquired TKI resistance, relapse or disease progression, which also remains a limitation of CML treatment. Understanding LSC resistance to TKI and development of strategies to increase the proportion of patients that may be cured of their leukemia is an exceptionally important area of investigation in CML.

A small subset of CML patients can maintain remission after treatment discontinuation [10]. However, even in patients maintaining TFR there is evidence that small numbers of leukemic cells persist long-term, suggesting that microenvironmental factors could play a role in determining the capability of residual LSC to regenerate leukemia after TKI discontinuation [11]. The current state of knowledge regarding the role of intrinsic alterations in gene regulation and of signals from the bone marrow microenvironment (BMM) in regulation of LSC,and their contribution to LSC persistence after treatment, and regeneration after stopping treatment, will be reviewed in this chapter.

6.2 Bone Marrow Microenvironmental Regulation of Normal Hematopoiesis

The hematopoietic system is a dynamic organ with over 100 billion cells bring replaced constantly. Hematopoietic stem cells (HSCs) are multipotent stem cells that are long-lived and able to self-renew as well as differentiate to mature lineages and replenish this pool of blood cells. HSCs reside in the bone marrow microenvironment and are located to specialized niches. For the most part, HSC remain quiescent, but can be released from quiescence as required for hematopoietic regeneration. There has been a considerable interest in understanding the role of the bone marrow microenvironment in regulating HSCs and how these signals are perturbed in hematological malignancies (Fig. 6.1).

The niche hypothesis proposes that stem cells reside in specialized locations in the microenvironment that protect HSCs and provide signals to modulate their fate [12]. Cells potentially providing niche function for HSC include mesenchymal stromal cells, endothelial cells, neuronal cells, osteoblasts, and hematopoietic cells such as macrophages, megakaryocytes and osteoclasts [13, 14]. These cells types lie close to the HSCs and provide necessary signals to modulate HSC

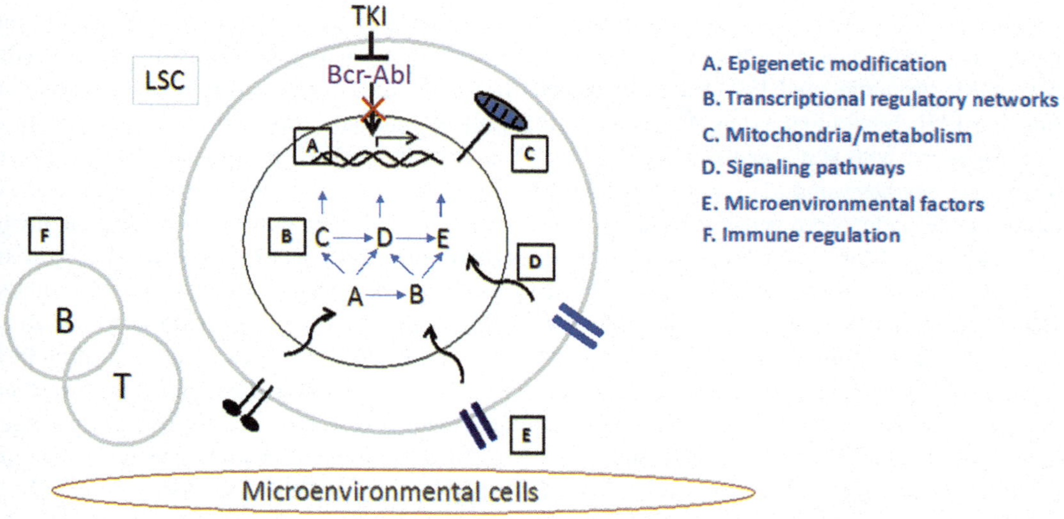

Fig. 6.1 Mechanisms underlying persistence of chronic myelogenous leukemia (CML) leukemia stem cells (LSC). Persistence of LSC after tyrosine kinase inhibitor (TKI) treatment is related to tyrosine kinase independent mechanisms which include intrinsic properties of LSCs determined by epigenetic alterations, altered transcriptional regulatory networks or mitochondrial/metabolic changes. These in turn are influenced by altered cell signaling, driven by alterations in microenvironmental interactions. Finally, these intrinsic alterations may also contribute to the ability of leukemia stem cells to escape immune regulation. Each type of alteration may offer potential points of intervention for therapeutic targeting of LSC

self-renewal, quiescence, differentiation and migration function.

6.2.1 Osteoblasts

Osteoblasts are bone-forming progenitor cells located at the surface of the endosteum that work in tandem with osteoclasts (bone-resorbing cells) to regulate osteogenesis [14]. Parathyroid hormone stimulation of osteoblastic activity has been reported to increase the HSC pool by expansion of the endosteal niche [15]. Visnjic et al. showed that osteoblast ablation leads to loss of BM cellularity, decreased numbers of HSC in the BM, and increased extramedullary hematopoiesis [16]. Osteoblasts in trabecular bone expressing high levels of Jagged-1 supported a high frequency of HSC. Increase in numbers of spindle-shaped N-cadherin+ osteoblasts following conditional inactivation of BMP was associated with an increase in HSC numbers [17]. Osteoblasts modulate HSC via several signaling pathways, including interaction of Jagged-1 with Notch receptors on HSCs, and of Angiopoietin (Ang-1) with Tie-2 (receptor tyrosine kinase) on HSCs [16, 18]. Osteopontin, a negative regulator of HSC proliferation, expressed on osteoblasts affects the expression of Jag-1 and Ang-1 on stromal cells. Loss of osteopontin enhanced HSC cycling and promoted HSC apoptosis [19]. On the other hand, deletion of CXCL12 and SCF, which are essential for HSC localization and maintenance from osteoblasts, did not alter HSC homeostasis [20, 21]. These studies exclude a role for OB expression of these specific factors in HSC regulation, but do not exclude other OB mediated mechanisms in HSC regulation.

6.2.2 Mesenchymal Stromal Stem Cells (MSCs)

MSCs are stromal progenitors that can give rise to bone, cartilage, fat, and fibroblast-like cells in the bone marrow. MSCs are located at vascular niches, in close proximity to arteriolar, transitional and sinusoidal vessels. The identification of MSCs has been challenging due to heterogeneity in function and markers. However, using genetic targeting strategies, cells expressing Nestin, the leptin-receptor or Prx1 been shown to have MSC properties and HSC niche regulator function [22]. Nestin-expressing MSCs showed co-localization with HSCs and adrengenic nerve fibers [23]. Studies using transgenic Nestin-GFP mouse model indicated a peri-arteriolar location for Nestin-dim MSCs near the endosteal bone surface, whereas Nestin-bright MSCs were located more centrally in the bone marrow in a peri-sinusoidal location [24]. Deletion of nestin-positive MSCs significantly reduced HSC in the bone marrow and increased HSC mobilization in vivo [22]. Loss of CXCL12 and SCF from either Prx1+ or LepR+ MSCs resulted in loss of HSC numbers and function [20, 21]. Therefore, MSC play a major role in regulating HSC homing, quiescence and differentiation.

6.2.3 CXCL12–Abundant Reticular (CAR) Cells

CXCL12 (SDF-1) is a chemokine elaborated by bone marrow stromal cells. It is a chemoattractant and binds to CXCR4, a seven-transmembrane G-coupled receptor protein which are expressed on HSCs. CXCL12-CXCR4 signaling is involved in HSC homing to the bone marrow and modulating HSC survival through regulation of integrin function [25]. While CXCL12 is expressed to an extent on endothelial and osteoblasts, CAR cells express high levels of CXCL12, have MSC properties, and do not express endothelial and osteoblast markers. CAR cells localize throughout the marrow at both the endosteal and sinusoidal niche areas. CAR cells have been shown to be important for B cell precursor development and HSC homing and their In vivo ablation severely impaired adipogenic and osteogenic differentiation together with a decrease in B cell and increase in HSC quiescence [26].

6.2.4 Endothelial Cells

Endothelial cells line blood vessels within the bone marrow microenvironment. The bone marrow vasculature supports low rates of blood flow, which creates relatively hypoxic conditions in the microenvironment. Endothelial cells are not only involved in the generation of vascular and extravascular space but also regulate HSCs. In conditions of stress, infection or bleeding, HSCs migrate to the vascular niche where they proliferate and differentiation in response to signals from endothelial cells. Endothelial cells secrete growth factors such as insulin-like growth factor binding protein-3 (IGFB3) that induce HSC proliferation. They also express CXCL12 which plays an important role in HSC and hematopoietic progenitor homing [21]. Loss of SCF from endothelial cells in mice results in a marked reduction of HSC frequency [27]. Notch ligands and E-selectins are also express on endothelial cells that bind and modulate HSC dormancy [28].

6.2.5 Neuronal Cells

The sympathetic nervous system regulates the attraction of HSC to their niche. UDP-galactose ceramide galactosyltransferase-deficient (Cgt−/−) mice demonstrated increased nerve conduction, causing a suppression of HSC egress during G-SCF administration. The defect was due to suppression of osteoblasts and decreased CXCL12 synthesis [23]. These cells have been shown to activate latent TGF-beta produced by a variety of cells. Autonomic nerve denervation reduced the number of active TGF-beta producing cells, consequently reducing the number of HSCs, suggesting that neuronal cells help maintain HSC through the activation of latent TGF-beta [29]. This suggests that the autonomic nervous system provides neuroprotection to the HSC and its niche. Indeed, chemotherapy-induced nerve injury impaired hematopoietic regeneration further indicating that nervous system plays an important role in protecting the HSCs within the bone marrow niche [30].

6.2.6 Hematopoietic Cells within the Niche

Several cells of hematopoietic origin may contribute to HSC niche regulation. Osteoclasts are bone-resorbing cells involved in endochondral ossification. Osteoclasts have been hypothesized to contribute to the endosteal niche by coordinating with osteoblasts in HSC regulation [14, 31]. Inhibition of osteoclastic activity increased total bone mass within the bone marrow and expanded the HSC pool in the endosteal niche and reduced HSC mobilization [31]. Treatment with bisphosphonates inhibited osteoclastic function and reduced the number of HSC within the endosteal region, potentially by disruption of the coupling of osteoclasts and osteoblasts [32]. Osteoclast activity increases the local and systemic calcium ion concentration, acting upon calcium-sensing receptors (CaR) expressed on HSCs. Mice deficient in CaR showed depressed bone marrow cellularity and HSC and increased mobilization of hematopoietic progenitor cells [33].

Immune cells have also been shown to regulate HSCs, and drive HSC proliferation and/or differentiation in response to infection and inflammation. An immune-privileged niche site may protect HSCs from autoimmunity, excessive inflammation and apoptosis. HSPCs colocalize with Tregs on the endosteal surface in the calvarial and trabecular BM, which may provide such a protective mechanism. A Treg subpopulation expressing high levels of CD150 localized within the HSC niche was found to control HSC quiescence and engraftment [34, 35].

Other hematopoietic-origin cells in the BM microenvironment may also influence niche function of MSC. For example, depletion of CD169+ macrophages in the bone marrow resulted in downregulation of HSC retention genes in the Nestin+ MSCs, including downregulation of CXCL12, and egress of HSCs into the bloodstream [36]. Megakaryocytes may also provide niche support for HSC quiescence through CXCL4 expression.[37] FGF1 production by megakaryocytes can also promote HSC expansion under stress.

6.3 Leukemia Stem Cell Resistance and Regulation by the Microenvironment in CML

While hematopoiesis is tightly regulated within the bone marrow, it is not surprising that perturbed signals and genetic insults can cause hematopoietic cells to become malignant and cancerous. Myeloid leukemias have been shown to arise from a leukemic stem cell via clonal expansion. LSCs share features common to HSCs including cell quiescence, self-renewal and differentiation. LSCs, however, differ with HSCs in activating signaling pathways for proliferation, invasion and survival. In CML the BCR-ABL translocation results in a constitutive activation of the ABL1 kinase and key downstream signaling pathways including JAK/STAT, Ras/MAPK and PI-3 K, that increases proliferation of malignant cells leading to development of a myeloproliferative disorder [1].

Specialized BM niches play a critical role in regulating HSC fate. However, the role of BM niches in regulating LSC fate is not well understood. Genetic lesions in leukemia cells provide cell-autonomous growth signals to leukemic cells but also alter how LSC interact with and respond to the BMM. It has long been known that CML progenitors demonstrate reduced adhesion to stroma [38]. LSCs may also show altered dependency on BMM signals. For example, LSCs in murine models of AML are insensitive to TGF-beta-mediated inhibition and show reduced dependence on Wnt signals compared to normal HSCs [39, 40]. In addition, LSC homing may be more dependent on CD44 and selectins compared to normal HSCs [41, 42].

It is believed that the BMM provides a sanctuary for LSC giving them clonal advantage. Colmone et al. showed that when murine leukemic cells were transplanted, they initially migrated towards CXCL12 enriched vascular niche and during leukemic growth, the CXCL12 production was markedly downregulated [43]. Other studies have shown that AML LSCs recruit Tregs to the site allowing them to escape detection by the immune cells [44]. Therefore, it is thought that LSCs hijack HSC niches where they are provided with survival signals through cell adhesion and 're-model' the niche to outcompete with HSCs. Indeed, leukemic patients with relatively low tumor burden have suppression of normal hematopoiesis. This may reflect the disruption of normal HSC and their BM niche.

We and others have reported that development of CML is associated with markedly altered levels of inflammatory cytokines and chemokines within the BM. These BMM alterations provide CML LSC with a growth advantage over normal HSC [45]. It is increasingly recognized that leukemic cells can induce changes in the BMM to enhance support of leukemic hematopoiesis at the expense of normal hematopoiesis. LSC can significantly remodeling the BMM disrupting normal HSC-BMM interactions [43, 45, 46]. Indeed, an abnormal BMM can even induce the development of myeloid leukemias by itself by inducing hematopoietic dysfunction [47–51]. These results are supported by the observation of secondary leukemias that originate from donor cells in rare patients after allogeneic bone marrow transplantation.

6.3.1 Adhesion Molecules

Adhesion molecules are primarily expressed on leukocytes and HSC to mediate migration between the bone marrow and other extramedullary sites. Very late antigen-4 (VLA-4), a heterodimer comprised of α4 and β1 subunits, is a member of the integrin family. HSCs expressing VLA-4 bind to osteoblasts and endothelial cells via the vascular cell adhesion molecule 1 (VCAM-1). VLA-4 is variably expressed on AML blasts and low expression of this molecule has been associated with increased risk of relapse in AML [52]. The bone marrow niche is rich in hyaluronic acid. CD44, a transmembrane glycoprotein, is a receptor for hyaluronic acid playing an important role in cell adhesion, homing and migration. BCR-ABL-expressing leukemic stem cells depend to a greater extent on CD44 for homing and engraftment than do normal HSCs, and argue that CD44 blockade may be beneficial in autologous transplantation in CML [41].

The CXCR4-CXCL12 axis plays a key role in HSC migration and homing to the bone marrow niche. LSCs have been found to exploit the CXCR4-CXCL12 axis in hematological malignancies. CML development decreased the levels of CXCL12 in the bone marrow. Loss of CXCL12 from the bone marrow led to an expansion of CML LSCs and increased migration to the spleen [45]. CXCR4 expressed on AML LSCs plays a pivotal role in anchoring immature LSCs in the bone marrow, maintaining quiescence [53]. AML cells upregulated expression of CXCR4 upon chemotherapy and increased stroma-mediated survival.

6.3.2 Wnt Signaling

Wnts are a family of secreted glycoproteins that interact with membrane-associated receptors to activate signaling cascades. Wnt-signaling is important for embryonic development, cell differentiation, proliferation and polarity. Canonical Wnt signaling is mediated by β-catenin stabilization and translocation into the nucleus, complex formation with TCF/LEF, and transcription of target genes. The effects of Wnt signaling are modulated by several factors including developmental stage, Wnt dosage, and microenvironmental factors. Aberrant activation of Wnt signaling is reported in several hematologic malignancies. Studies combining different *APC* conditional knockouts to create five different levels of Wnt signaling *in vivo* indicated that levels of Wnt signaling affect the balance of HSC self-renewal versus differentiation [54]. A decrease in Wnt signaling leads to loss of HSC quiescence and self-renewal, mild increases in Wnt activity supports increased hematopoietic reconstitution, and high levels of Wnt activation impair HSC self-renewal and differentiation.

Enhanced Wnt-β-catenin activity may contribute to LSC transformation in blast crisis (BC) CML [55, 56]. BCR-ABL-expressing β-catenin deleted HSC demonstrated reduced LSC self-renewal and leukemia development [57, 58].

β-catenin deletion in established leukemia synergized with IM to eliminate CML stem cells [59]. Increased β-catenin in CML cells results from BCR-ABL-mediated β-catenin phosphorylation [60], and GSK3β inactivation downstream of BCR-ABL or GSK3β mis-splicing leading to reduced β-catenin degradation [61] [60]. N-Cadherin-mediated adhesion of CML LSC to mesenchymal cells enhanced β-catenin nuclear translocation and transcriptional activity and protected CML progenitors from TKI treatment. Although Wnt signaling is recognized as a key regulator of HSC and LSC fate, the complexity of this regulatory mechanism, and the role and interaction between canonical versus non-canonical pathways in HSC needs to be better understood.

6.3.3 Hedgehog Pathway

Hedgehog (Hh) signaling involves three Hh ligand, Indian Hh (IHH), Sonic Hh (SHH), and Desert Hh (DHH), the smoothened (SMO) receptor, the transmembrane protein patched (PTCH) which is a negative regulator of SMO, and glioma zinc finger transcription factors (GLI1, GLI2, and GLI3). IHH is produced in the BM but is dispensable for normal hematopoiesis [62, 63]. In CML, Hh signaling is substantially increased in leukemic stem and progenitor cells as the disease progresses from chronic phase to accelerated and blast crisis [64–66]. Inhibition of Hh signaling reduced the number of LSC and decreased the development of leukemia. Mice treated with pharmacological inhibitors of Hh showed lower LSC activity and increased survival.

6.3.4 Notch Pathway

Notch signaling is a highly conserved mechanism that regulates many developmental decisions such as cell fate, proliferation, and differentiation. The central components of the Notch pathway include the four Notch receptors (Notch1–4) and five cognate ligands Delta-like (Dll1, Dll3 and Dll4), Jagged (Jag1, Jag2). Notch ligands are secretory proteins which require palmitoylation for extracellular secretion from the cell where they interact with the respective receptors to initiate downstream signaling [67]. The role of the Notch pathway has been controversial since the effects of Notch differ depending on cell conditions and are modified by other stemness-related signals. Notch activation can lead to quiescence of stem cells in the context of the BMM whereas in other scenarios it may be associated with loss of quiescence and increased proliferative state.

Osteoblasts in trabecular bone express high levels of Jagged-1, which interact with Notch receptors on HSCs. Jagged-1 inhibits HSC entry into cell cycle, and loss of Jagged-1 induced signaling following osteoblast ablation may contribute to loss of HSC quiescence. Jagged-1 was overexpressed in OB from BCR-ABL compared to normal mice. As with normal HSC, Jagged-1 exposure inhibited cell cycling of CML LSC, suggesting that loss of Jagged-1 signaling. Osteoblast ablation in a CML mouse model resulted in accelerated leukemia development and reduced survival. Osteoblast-expressed Jagged-1 may contribute to osteoblast mediated HSC and LSC regulation [68]. Further investigations of the role of Notch pathway in the regulation of HSC and LSC by the BMM are warranted.

6.4 Approaches to Targeting Dormant Leukemia Stem Cells

Improving understanding of mechanisms of LSC maintenance, and the complex cross-talk between LSC and the BMM is yielding opportunities for development of novel therapeutic strategies aimed at targeting this resistant population to improve outcomes of leukemia patients.

6.4.1 SIRT1, p53, and MYC

The p53 tumor suppressor is an important regulator of cell cycle and apoptosis. Activation of p53 using small molecule inhibitors of p53 interactions with its negative regulator HDM2 in CML appears to be effective in targeting CML LSCs both in vitro and in vivo [69]. The NAD+ dependent deacetylase SIRT1 is overexpressed in human CML LSC and inhibits p53 via deacetylation. Pharmacological inhibition of SIRT1 increases apoptosis and depletes CP CML LSC and may represent a potential approach to activate p53 to target CML LSC [70]. An unbiased analysis using proteomics, transcriptomics and network analyses showed that p53 and c-MYC represent critical, connected nodes that regulate human CML LSCs. The combination of MDM2 and BET inhibitors to upregulate p53 and downregulate c-MYC resulted in synergistic killing of human LSCs, while sparing normal HSCs [71].

6.4.2 Epigenetic Targeting

The HDACi LBH589 (Panabinostat) in combination with imatinib was effective in inducing apoptosis and eliminating CML LSC in vivo. The combination inhibited expression of genes regulating HSC maintenance and survival [72]. The Polycomb Repressive Complex 2 (PRC2) is dysregulated in CP CML LSC, with extensive reprogramming of H3K27me3 targets. Treatment with an EZH2 inhibitor (EZH2i) sensitized LSCs to apoptosis, without affecting normal HSC survival, and combined treatment with TKI enhanced depletion of LSCs compared to TKI alone. These observations support EZH2 inhibition as a strategy to effectively target LSCs in patients with CML receiving TKIs [73].

6.4.3 JAK2 Kinase Inhibition

The JAK kinases are intracellular nonreceptor kinases that mediate cytokine-mediated signaling via activation of STAT transcription factors. CML cells demonstrate increased STAT5 phosphorylation, nuclear translocation and transcriptional activity. STAT5 inactivation attenuates CML development in mouse models. Although BCR-ABL directly activates STAT5, JAK2 is also an upstream mediator of STAT signaling in CML LSC related to signaling through cytokines expressed in the BMM, and JAK2inhibition by ruxolitinib in combination with BCR-ABL TKIs can deplete CML LSCs [74]. Another strategy to inhibit JAK-STAT signaling in CML LSC is by activation of the serine-threonine phosphatase PP2A which is inactivated in CML LSCs. Reactivation of PP2A marked inhibits STAT5 and β-catenin signaling and depletes LSC [75].

6.4.4 Inflammatory Signaling

As discussed above, CML development is associated with altered function of the BMM, contributing to altered regulation of leukemic and normal stem cells. Leukemia development markedly alters inflammatory cytokine expression in the BM, selective impairs normal LTHSC growth, and provides a growth advantage to CML LSC. TKI treatment only partially corrects these abnormalities. Autocrine production of tumor necrosis factor-α (TNF-α) enhances CML LSC survival by promoting NFκB/p65 activity. Inhibition of TNF-α signaling via a small-molecule TNF-α inhibitor induced apoptosis in CML LSC, and in combination with TKI induces significantly more apoptosis and reduction in CML LSC than TKI alone [76]. Interleukin-1 (IL-1) is a pivotal pro-inflammatory cytokine that is overexpressed in CML BM, and its receptors IL-1 receptor accessory protein (IL1-RAP) and IL-1 receptor type 1, are upregulated on CML LSC, which demonstrate increased IL-1-induced signaling. Inhibition of IL-1 signaling with recombinant IL-1 receptor antagonist (IL-1RA) treatment inhibited growth of CML LSC, and the combination of IL-1RA with TKI resulted in significantly greater CML LSC inhibition than TKI alone [77]. Administration of IL1RAP antibodies enhanced elimination of human CML cells [78].

6.4.5 Developmental Factors

Secretion of Wnt ligands requires their modification by the O-acyl transferase Porcupine

(PORCN). Treatment with WNT974, a potent and selective PORCN inhibitor, antagonized Wnt signaling in human CML CD34+ cells, and significantly inhibited growth of CML LSC in combination with the TKI, compared with TKI alone [68]. Inhibition of Hedgehog signaling with pharmacological inhibition of SMO reduces the number and self-renewal capacity of CML LSC in vitro and their engraftment in NSG mice [79].

6.4.6 Stem Cell Mobilization

Plerixafor is a bicyclam small molecule inhibitor of CXCR4 that is approved for the mobilization of HSC for transplantation [16, 17]. CXCR4 blockade disrupts the ability of the BMM to protect AML cells from chemotherapy or TKI treatment in mice, decreasing the leukemia burden and improving overall survival [18]. CXCR4 blockade also suppresses stroma-activated survival pathways in AML cells. Several CXCR4 inhibitors are in clinical development either for stem cell mobilization, and could be of potential benefit in sensitizing CML LSC to TKI treatment.

Integrin-mediated interactions of LSC with BMM cells have been disrupted using a VLA-4 antibody, which was able to overcome resistance to chemotherapy in leukemia cells [33]. Alternative agents that block VLA-4 are in clinical development. A monoclonal antibody directed against CD44 to immune-deficient mice transplanted with human AML markedly reduced leukemic repopulation [80].

6.4.7 BCL-2 Inhibition

Increased BCL-2 protein expression is observed in CML BM cells. Selective inhibition of BCL-2, in combination with TKI, markedly decreased LSC cell numbers and prolonged survival in a murine CML model, and effectively eradicated LSC from CML patient samples. These results support the utility of combined inhibition of BCL-2 and BCR-ABL to target CML LSC [81].

6.5 Conclusion

Leukemic and normal stem cells share properties of quiescence and self-renewal, that are supported by none marrow niches. Our ability to identify and characterize LSC populations that persist in patients in molecular remission after TKI treatment is limited because of the rarity of these cells and lack of specific markers, curtailing our ability to isolate and directly characterize these cells. In addition, current xenograft models for in vivo analysis of CML LSC populations are inefficient. Recent advances in single cell analysis and cell tracking which may enhance our ability to characterize residual LSC. Identification of differences in cellular and molecular regulation of normal and leukemic stem cells remains critical to effective targeting of dormant disease regenerating leukemic populations in the future.

References

1. Sawyers CL (1999) Chronic myeloid leukemia. N Engl J Med 340(17):1330–1340. PubMed PMID: 10219069
2. Druker BJ, Guilhot F, O'Brien SG, Gathmann I, Kantarjian H, Gattermann N, Deininger MW, Silver RT, Goldman JM, Stone RM, Cervantes F, Hochhaus A, Powell BL, Gabrilove JL, Rousselot P, Reiffers J, Cornelissen JJ, Hughes T, Agis H, Fischer T, Verhoef G, Shepherd J, Saglio G, Gratwohl A, Nielsen JL, Radich JP, Simonsson B, Taylor K, Baccarani M, So C, Letvak L, Larson RA (2006) Five-year follow-up of patients receiving imatinib for chronic myeloid leukemia. N Engl J Med 355(23):2408–2417. PubMed PMID: 17151364
3. Chu S, McDonald T, Lin A, Chakraborty S, Huang Q, Snyder DS, Bhatia R (2011) Persistence of leukemia stem cells in chronic myelogenous leukemia patients in prolonged remission with imatinib treatment. Blood. Epub 2011/09/21. doi:blood-2010-12-327437 [pii] 10.1182/blood-2010-12-327437. PubMed PMID: 21931114
4. Chu S, Holtz M, Gupta M, Bhatia R (2004) BCR/ABL kinase inhibition by imatinib mesylate enhances MAP kinase activity in chronic myelogenous leukemia CD34+ cells. Blood 103(8):3167–3174. PubMed PMID: 15070699
5. Corbin AS, Agarwal A, Loriaux M, Cortes J, Deininger MW, Druker BJ (2011) Human chronic myeloid leukemia stem cells are insensitive to imatinib despite inhibition of BCR-ABL activity. J Clin Invest 121(1):396–409. PubMed PMID: 21157039

6. Graham SM, Jorgensen HG, Allan E, Pearson C, Alcorn MJ, Richmond L, Holyoake TL (2002) Primitive, quiescent, Philadelphia-positive stem cells from patients with chronic myeloid leukemia are insensitive to STI571 in vitro. Blood 99(1):319–325. PubMed PMID: 11756187
7. Holtz MS, Slovak ML, Zhang F, Sawyers CL, Forman SJ, Bhatia R (2002) Imatinib mesylate (STI571) inhibits growth of primitive malignant progenitors in chronic myelogenous leukemia through reversal of abnormally increased proliferation. Blood 99(10):3792–3800. PubMed PMID: 11986238
8. Zhang B, Li M, McDonald T, Holyoake TL, Moon RT, Campana D, Shultz L, Bhatia R (2013) Microenvironmental protection of CML stem and progenitor cells from tyrosine kinase inhibitors through N-cadherin and Wnt-beta-catenin signaling. Blood 121(10):1824–1838. https://doi.org/10.1182/blood-2012-02-412890. PubMed PMID: 23299311; PMCID: PMC3591802
9. Bhatia R (2017) Novel approaches to therapy in CML. Hematol Am Soc Hematol Educ Program 2017(1):115–120. Epub 2017/12/10. https://doi.org/10.1182/asheducation-2017.1.115. PubMed PMID: 29222245
10. Mahon FX, Rea D, Guilhot J, Guilhot F, Huguet F, Nicolini F, Legros L, Charbonnier A, Guerci A, Varet B, Etienne G, Reiffers J, Rousselot P (2010) Discontinuation of imatinib in patients with chronic myeloid leukaemia who have maintained complete molecular remission for at least 2 years: the prospective, multicentre Stop Imatinib (STIM) trial. Lancet Oncol 11(11):1029–1035. PubMed PMID: 20965785
11. Ross DM, Branford S, Seymour JF, Schwarer AP, Arthur C, Yeung DT, Dang P, Goyne JM, Slader C, Filshie RJ, Mills AK, Melo JV, White DL, Grigg AP, Hughes TP (2013) Safety and efficacy of imatinib cessation for CML patients with stable undetectable minimal residual disease: results from the TWISTER study. Blood 122(4):515–522. https://doi.org/10.1182/blood-2013-02-483750. PubMed PMID: 23704092
12. Schofield R (1978) The relationship between the spleen colony-forming cell and the haemopoietic stem cell. Blood Cells 4(1-2):7–25. Epub 1978/01/01. PubMed PMID: 747780
13. Morrison SJ, Scadden DT (2014) The bone marrow niche for haematopoietic stem cells. Nature 505(7483):327–334. https://doi.org/10.1038/nature12984. PubMed PMID: 24429631; PMCID: PMC4514480
14. Mansour A, Abou-Ezzi G, Sitnicka E, Jacobsen SE, Wakkach A, Blin-Wakkach C (2012) Osteoclasts promote the formation of hematopoietic stem cell niches in the bone marrow. J Exp Med 209(3):537–549. Epub 2012/02/22. https://doi.org/10.1084/jem.20110994. PubMed PMID: 22351931; PMCID: PMC3302238
15. Visnjic D, Kalajzic Z, Rowe DW, Katavic V, Lorenzo J, Aguila HL (2004) Hematopoiesis is severely altered in mice with an induced osteoblast deficiency. Blood 103(9):3258–3264. Epub 2004/01/17. https://doi.org/10.1182/blood-2003-11-4011. PubMed PMID: 14726388
16. Calvi LM, Adams GB, Weibrecht KW, Weber JM, Olson DP, Knight MC, Martin RP, Schipani E, Divieti P, Bringhurst FR, Milner LA, Kronenberg HM, Scadden DT (2003) Osteoblastic cells regulate the haematopoietic stem cell niche. Nature 425(6960):841–846. Epub 2003/10/24. https://doi.org/10.1038/nature02040. PubMed PMID: 14574413
17. Zhang J, Niu C, Ye L, Huang H, He X, Tong WG, Ross J, Haug J, Johnson T, Feng JQ, Harris S, Wiedemann LM, Mishina Y, Li L (2003) Identification of the haematopoietic stem cell niche and control of the niche size. Nature 425(6960):836–841. Epub 2003/10/24. https://doi.org/10.1038/nature02041. PubMed PMID: 14574412
18. Arai F, Hirao A, Ohmura M, Sato H, Matsuoka S, Takubo K, Ito K, Koh GY, Suda T (2004) Tie2/angiopoietin-1 signaling regulates hematopoietic stem cell quiescence in the bone marrow niche. Cell 118(2):149–161. Epub 2004/07/21. https://doi.org/10.1016/j.cell.2004.07.004. PubMed PMID: 15260986
19. Nilsson SK, Johnston HM, Whitty GA, Williams B, Webb RJ, Denhardt DT, Bertoncello I, Bendall LJ, Simmons PJ, Haylock DN (2005) Osteopontin, a key component of the hematopoietic stem cell niche and regulator of primitive hematopoietic progenitor cells. Blood 106(4):1232–1239. Epub 2005/04/23. https://doi.org/10.1182/blood-2004-11-4422. PubMed PMID: 15845900
20. Greenbaum A, Hsu YM, Day RB, Schuettpelz LG, Christopher MJ, Borgerding JN, Nagasawa T, Link DC (2013) CXCL12 in early mesenchymal progenitors is required for haematopoietic stem-cell maintenance. Nature 495(7440):227–230. https://doi.org/10.1038/nature11926. PubMed PMID: 23434756; PMCID: PMC3600148
21. Ding L, Morrison SJ (2013) Haematopoietic stem cells and early lymphoid progenitors occupy distinct bone marrow niches. Nature 495(7440):231–235. https://doi.org/10.1038/nature11885. PubMed PMID: 23434755; PMCID: PMC3600153
22. Mendez-Ferrer S, Michurina TV, Ferraro F, Mazloom AR, Macarthur BD, Lira SA, Scadden DT, Ma'ayan A, Enikolopov GN, Frenette PS (2010) Mesenchymal and haematopoietic stem cells form a unique bone marrow niche. Nature 466(7308):829–834. Epub 2010/08/13. https://doi.org/10.1038/nature09262. PubMed PMID: 20703299; PMCID: PMC3146551
23. Katayama Y, Battista M, Kao WM, Hidalgo A, Peired AJ, Thomas SA, Frenette PS (2006) Signals from the sympathetic nervous system regulate hematopoietic stem cell egress from bone marrow. Cell 124(2):407–421. Epub 2006/01/28. https://doi.org/10.1016/j.cell.2005.10.041. PubMed PMID: 16439213
24. Kunisaki Y, Bruns I, Scheiermann C, Ahmed J, Pinho S, Zhang D, Mizoguchi T, Wei Q, Lucas D, Ito K, Mar JC, Bergman A, Frenette PS (2013) Arteriolar niches

maintain haematopoietic stem cell quiescence. Nature 502(7473):637–643. Epub 2013/10/11. https://doi.org/10.1038/nature12612. PubMed PMID: 24107994; PMCID: PMC3821873
25. Sugiyama T, Kohara H, Noda M, Nagasawa T (2006) Maintenance of the hematopoietic stem cell pool by CXCL12-CXCR4 chemokine signaling in bone marrow stromal cell niches. Immunity 25(6):977–988. Epub 2006/12/19. doi:S1074-7613(06)00515-2 [pii]10.1016/j.immuni.2006.10.016. PubMed PMID: 17174120
26. Omatsu Y, Sugiyama T, Kohara H, Kondoh G, Fujii N, Kohno K, Nagasawa T (2010) The essential functions of adipo-osteogenic progenitors as the hematopoietic stem and progenitor cell niche. Immunity 33(3):387–399. Epub 2010/09/21. https://doi.org/10.1016/j.immuni.2010.08.017. PubMed PMID: 20850355
27. Ding L, Saunders TL, Enikolopov G, Morrison SJ (2012) Endothelial and perivascular cells maintain haematopoietic stem cells. Nature 481(7382):457–462. https://doi.org/10.1038/nature10783. PubMed PMID: 22281595; PMCID: PMC3270376
28. Winkler IG, Barbier V, Nowlan B, Jacobsen RN, Forristal CE, Patton JT, Magnani JL, Levesque JP (2012) Vascular niche E-selectin regulates hematopoietic stem cell dormancy, self renewal and chemoresistance. Nat Med 18(11):1651–1657. Epub 2012/10/23. https://doi.org/10.1038/nm.2969. PubMed PMID: 23086476
29. Yamazaki S, Ema H, Karlsson G, Yamaguchi T, Miyoshi H, Shioda S, Taketo MM, Karlsson S, Iwama A, Nakauchi H (2011) Nonmyelinating Schwann cells maintain hematopoietic stem cell hibernation in the bone marrow niche. Cell 147(5):1146–1158. Epub 2011/11/29. https://doi.org/10.1016/j.cell.2011.09.053. PubMed PMID: 22118468
30. Lucas D, Scheiermann C, Chow A, Kunisaki Y, Bruns I, Barrick C, Tessarollo L, Frenette PS (2013) Chemotherapy-induced bone marrow nerve injury impairs hematopoietic regeneration. Nat Med 19(6):695–703. Epub 2013/05/07. https://doi.org/10.1038/nm.3155. PubMed PMID: 23644514; PMCID: PMC3964478
31. Miyamoto T (2013) Role of osteoclasts in regulating hematopoietic stem and progenitor cells. World J Orthop 4(4):198–206. Epub 2013/10/23. https://doi.org/10.5312/wjo.v4.i4.198. PubMed PMID: 24147255; PMCID: PMC3801239
32. Lymperi S, Ersek A, Ferraro F, Dazzi F, Horwood NJ (2011) Inhibition of osteoclast function reduces hematopoietic stem cell numbers in vivo. Blood 117(5):1540–1549. Epub 2010/12/07. https://doi.org/10.1182/blood-2010-05-282855. PubMed PMID: 21131587
33. Adams GB, Martin RP, Alley IR, Chabner KT, Cohen KS, Calvi LM, Kronenberg HM, Scadden DT (2007) Therapeutic targeting of a stem cell niche. Nat Biotechnol 25(2):238–243. Epub 2007/01/24. https://doi.org/10.1038/nbt1281. PubMed PMID: 17237769
34. Fujisaki J, Wu J, Carlson AL, Silberstein L, Putheti P, Larocca R, Gao W, Saito TI, Lo Celso C, Tsuyuzaki H, Sato T, Cote D, Sykes M, Strom TB, Scadden DT, Lin CP (2011) In vivo imaging of Treg cells providing immune privilege to the haematopoietic stem-cell niche. Nature 474(7350):216–219. Epub 2011/06/10. https://doi.org/10.1038/nature10160. PubMed PMID: 21654805; PMCID: PMC3725645
35. Hirata Y, Furuhashi K, Ishii H, Li HW, Pinho S, Ding L, Robson SC, Frenette PS, Fujisaki J (2018) CD150(high) bone marrow tregs maintain hematopoietic stem cell quiescence and immune privilege via adenosine. Cell Stem Cell 22(3):445–53 e5. Epub 2018/02/20. https://doi.org/10.1016/j.stem.2018.01.017. PubMed PMID: 29456159
36. Chow A, Lucas D, Hidalgo A, Mendez-Ferrer S, Hashimoto D, Scheiermann C, Battista M, Leboeuf M, Prophete C, van Rooijen N, Tanaka M, Merad M, Frenette PS (2011) Bone marrow CD169+ macrophages promote the retention of hematopoietic stem and progenitor cells in the mesenchymal stem cell niche. J Exp Med 208(2):261–271. Epub 2011/02/02. https://doi.org/10.1084/jem.20101688. PubMed PMID: 21282381; PMCID: PMC3039855
37. Bruns I, Lucas D, Pinho S, Ahmed J, Lambert MP, Kunisaki Y, Scheiermann C, Schiff L, Poncz M, Bergman A, Frenette PS (2014) Megakaryocytes regulate hematopoietic stem cell quiescence through CXCL4 secretion. Nat Med 20(11):1315–1320. Epub 2014/10/20. https://doi.org/10.1038/nm.3707. PubMed PMID: 25326802; PMCID: PMC4258871
38. Gordon MY, Dowding CR, Riley GP, Goldman JM, Greaves MF (1987) Altered adhesive interactions with marrow stroma of haematopoietic progenitor cells in chronic myeloid leukaemia. Nature 328(6128):342–344. Epub 1987/07/23. https://doi.org/10.1038/328342a0. PubMed PMID: 3474529
39. Krause DS, Fulzele K, Catic A, Sun CC, Dombkowski D, Hurley MP, Lezeau S, Attar E, Wu JY, Lin HY, Divieti-Pajevic P, Hasserjian RP, Schipani E, Van Etten RA, Scadden DT (2013) Differential regulation of myeloid leukemias by the bone marrow microenvironment. Nat Med 19(11):1513–1517. Epub 2013/10/29. https://doi.org/10.1038/nm.3364. PubMed PMID: 24162813; PMCID: PMC3827980
40. Lane SW, Wang YJ, Lo Celso C, Ragu C, Bullinger L, Sykes SM, Ferraro F, Shterental S, Lin CP, Gilliland DG, Scadden DT, Armstrong SA, Williams DA (2011) Differential niche and Wnt requirements during acute myeloid leukemia progression. Blood 118(10):2849–2856. Epub 2011/07/19. https://doi.org/10.1182/blood-2011-03-345165. PubMed PMID: 21765021; PMCID: PMC3172801
41. Krause DS, Lazarides K, von Andrian UH, Van Etten RA (2006) Requirement for CD44 in homing and engraftment of BCR-ABL-expressing leukemic stem cells. Nat Med 12(10):1175–1180. PubMed PMID: 16998483
42. Krause DS, Lazarides K, Lewis JB, von Andrian UH, Van Etten RA (2014) Selectins and their ligands are

required for homing and engraftment of BCR-ABL1+ leukemic stem cells in the bone marrow niche. Blood 123(9):1361–1371. Epub 2014/01/08. https://doi.org/10.1182/blood-2013-11-538694. PubMed PMID: 24394666; PMCID: PMC3938148
43. Colmone A, Amorim M, Pontier AL, Wang S, Jablonski E, Sipkins DA (2008) Leukemic cells create bone marrow niches that disrupt the behavior of normal hematopoietic progenitor cells. Science (New York, NY) 322(5909):1861–1865. Epub 2008/12/20. https://doi.org/10.1126/science.1164390. PubMed PMID: 19095944
44. Zhou Q, Bucher C, Munger ME, Highfill SL, Tolar J, Munn DH, Levine BL, Riddle M, June CH, Vallera DA, Weigel BJ, Blazar BR (2009) Depletion of endogenous tumor-associated regulatory T cells improves the efficacy of adoptive cytotoxic T-cell immunotherapy in murine acute myeloid leukemia. Blood 114(18):3793–3802. Epub 2009/09/03. https://doi.org/10.1182/blood-2009-03-208181. PubMed PMID: 19724059; PMCID: PMC2773484
45. Zhang B, Ho YW, Huang Q, Maeda T, Lin A, Lee SU, Hair A, Holyoake TL, Huettner C, Bhatia R (2012) Altered microenvironmental regulation of leukemic and normal stem cells in chronic myelogenous leukemia. Cancer Cell 21(4):577–592. https://doi.org/10.1016/j.ccr.2012.02.018. PubMed PMID: 22516264; PMCID: 3332001
46. Schepers K, Pietras EM, Reynaud D, Flach J, Binnewies M, Garg T, Wagers AJ, Hsiao EC, Passegue E (2013) Myeloproliferative neoplasia remodels the endosteal bone marrow niche into a self-reinforcing leukemic niche. Cell Stem Cell 13(3):285–299. https://doi.org/10.1016/j.stem.2013.06.009. PubMed PMID: 23850243; PMCID: 3769504
47. Shiozawa Y, Taichman RS (2010) Dysfunctional niches as a root of hematopoietic malignancy. Cell Stem Cell 6(5):399–400. https://doi.org/10.1016/j.stem.2010.04.004. PubMed PMID: 20452309
48. Raaijmakers MH, Mukherjee S, Guo S, Zhang S, Kobayashi T, Schoonmaker JA, Ebert BL, Al-Shahrour F, Hasserjian RP, Scadden EO, Aung Z, Matza M, Merkenschlager M, Lin C, Rommens JM, Scadden DT (2010) Bone progenitor dysfunction induces myelodysplasia and secondary leukaemia. Nature 464(7290):852–857. https://doi.org/10.1038/nature08851. PubMed PMID: 20305640; PMCID: 3422863
49. Walkley CR, Olsen GH, Dworkin S, Fabb SA, Swann J, McArthur GA, Westmoreland SV, Chambon P, Scadden DT, Purton LE (2007) A microenvironment-induced myeloproliferative syndrome caused by retinoic acid receptor gamma deficiency. Cell 129(6):1097–1110. https://doi.org/10.1016/j.cell.2007.05.014. PubMed PMID: 17574023; PMCID: 1974882
50. Walkley CR, Shea JM, Sims NA, Purton LE, Orkin SH (2007) Rb regulates interactions between hematopoietic stem cells and their bone marrow microenvironment. Cell 129(6):1081–1095. https://doi.org/10.1016/j.cell.2007.03.055. PubMed PMID: 17574022; PMCID: 2768301
51. Kim YW, Koo BK, Jeong HW, Yoon MJ, Song R, Shin J, Jeong DC, Kim SH, Kong YY (2008) Defective Notch activation in microenvironment leads to myeloproliferative disease. Blood 112(12):4628–4638. https://doi.org/10.1182/blood-2008-03-148999. PubMed PMID: 18818392
52. Matsunaga T, Takemoto N, Sato T, Takimoto R, Tanaka I, Fujimi A, Akiyama T, Kuroda H, Kawano Y, Kobune M, Kato J, Hirayama Y, Sakamaki S, Kohda K, Miyake K, Niitsu Y (2003) Interaction between leukemic-cell VLA-4 and stromal fibronectin is a decisive factor for minimal residual disease of acute myelogenous leukemia. Nat Med 9(9):1158–1165. Epub 2003/08/05. https://doi.org/10.1038/nm909. PubMed PMID: 12897778
53. Voermans C, van Heese WP, de Jong I, Gerritsen WR, van Der Schoot CE (2002) Migratory behavior of leukemic cells from acute myeloid leukemia patients. Leukemia 16(4):650–657. Epub 2002/04/18. https://doi.org/10.1038/sj.leu.2402431. PubMed PMID: 11960346
54. Luis TC, Naber BA, Roozen PP, Brugman MH, de Haas EF, Ghazvini M, Fibbe WE, van Dongen JJ, Fodde R, Staal FJ (2011) Canonical wnt signaling regulates hematopoiesis in a dosage-dependent fashion. Cell Stem Cell 9(4):345–356. Epub 2011/10/11. https://doi.org/10.1016/j.stem.2011.07.017. PubMed PMID: 21982234
55. Minami Y, Stuart SA, Ikawa T, Jiang Y, Banno A, Hunton IC, Young DJ, Naoe T, Murre C, Jamieson CH, Wang JY (2008) BCR-ABL-transformed GMP as myeloid leukemic stem cells. Proc Natl Acad Sci U S A 105(46):17967–17972. PubMed PMID: 19004799
56. Jamieson CH, Ailles LE, Dylla SJ, Muijtjens M, Jones C, Zehnder JL, Gotlib J, Li K, Manz MG, Keating A, Sawyers CL, Weissman IL (2004) Granulocyte-macrophage progenitors as candidate leukemic stem cells in blast-crisis CML. N Engl J Med 351(7):657–667. PubMed PMID: 15306667
57. Zhao C, Blum J, Chen A, Kwon HY, Jung SH, Cook JM, Lagoo A, Reya T (2007) Loss of beta-catenin impairs the renewal of normal and CML stem cells in vivo. Cancer Cell 12(6):528–541. Epub 2007/12/11. doi:S1535-6108(07)00334-0 [pii]10.1016/j.ccr.2007.11.003. PubMed PMID: 18068630; PMCID: 2262869.
58. Hu Y, Chen Y, Douglas L, Li S (2009) beta-Catenin is essential for survival of leukemic stem cells insensitive to kinase inhibition in mice with BCR-ABL-induced chronic myeloid leukemia. Leukemia 23(1):109–116. Epub 2008/09/27. https://doi.org/10.1038/leu.2008.262. PubMed PMID: 18818703
59. Heidel FH, Bullinger L, Feng Z, Wang Z, Neff TA, Stein L, Kalaitzidis D, Lane SW, Armstrong SA (2012) Genetic and pharmacologic inhibition of beta-catenin targets imatinib-resistant leukemia stem cells in CML. Cell Stem Cell 10(4):412–424. Epub 2012/04/10. doi:S1934-5909(12)00076-8

[pii]10.1016/j.stem.2012.02.017. PubMed PMID: 22482506; PMCID: 3339412

60. Coluccia AM, Vacca A, Dunach M, Mologni L, Redaelli S, Bustos VH, Benati D, Pinna LA, Gambacorti-Passerini C (2007) Bcr-Abl stabilizes beta-catenin in chronic myeloid leukemia through its tyrosine phosphorylation. EMBO J 26(5):1456–1466. Epub 2007/02/24. doi:7601485 [pii]10.1038/sj.emboj.7601485. PubMed PMID: 17318191; PMCID: 1817619
61. Samanta AK, Lin H, Sun T, Kantarjian H, Arlinghaus RB (2006) Janus kinase 2: a critical target in chronic myelogenous leukemia. Cancer Res 66(13):6468–6472. PubMed PMID: 16818614
62. Gao J, Graves S, Koch U, Liu S, Jankovic V, Buonamici S, El Andaloussi A, Nimer SD, Kee BL, Taichman R, Radtke F, Aifantis I (2009) Hedgehog signaling is dispensable for adult hematopoietic stem cell function. Cell Stem Cell 4(6):548–558. https://doi.org/10.1016/j.stem.2009.03.015. PubMed PMID: 19497283; PMCID: 2914688
63. Hofmann I, Stover EH, Cullen DE, Mao J, Morgan KJ, Lee BH, Kharas MG, Miller PG, Cornejo MG, Okabe R, Armstrong SA, Ghilardi N, Gould S, de Sauvage FJ, McMahon AP, Gilliland DG (2009) Hedgehog signaling is dispensable for adult murine hematopoietic stem cell function and hematopoiesis. Cell Stem Cell 4(6):559–567. https://doi.org/10.1016/j.stem.2009.03.016. PubMed PMID: 19497284; PMCID: 3065323
64. Dierks C, Beigi R, Guo GR, Zirlik K, Stegert MR, Manley P, Trussell C, Schmitt-Graeff A, Landwerlin K, Veelken H, Warmuth M (2008) Expansion of Bcr-Abl-positive leukemic stem cells is dependent on Hedgehog pathway activation. Cancer Cell 14(3):238–249. Epub 2008/09/06. doi:S1535-6108(08)00257-2 [pii]10.1016/j.ccr.2008.08.003. PubMed PMID: 18772113
65. Zhao C, Chen A, Jamieson CH, Fereshteh M, Abrahamsson A, Blum J, Kwon HY, Kim J, Chute JP, Rizzieri D, Munchhof M, VanArsdale T, Beachy PA, Reya T (2009) Hedgehog signalling is essential for maintenance of cancer stem cells in myeloid leukaemia. Nature 458(7239):776–779. https://doi.org/10.1038/nature07737. PubMed PMID: 19169242; PMCID: 2946231
66. Irvine DA, Copland M (2012) Targeting hedgehog in hematologic malignancy. Blood 119(10):2196–2204. https://doi.org/10.1182/blood-2011-10-383752. PubMed PMID: 22223823
67. Evans AG, Calvi LM (2015) Notch signaling in the malignant bone marrow microenvironment: implications for a niche-based model of oncogenesis. Ann N Y Acad Sci 1335(1):63–77. https://doi.org/10.1111/nyas.12562. PubMed PMID: 25351294; PMCID: 4289406
68. Agarwal P, Zhang B, Ho Y, Cook A, Li L, Mikhail FM, Wang Y, McLaughlin ME, Bhatia R (2017) Enhanced targeting of CML stem and progenitor cells by inhibition of porcupine acyltransferase in combination with TKI. Blood 129(8):1008–1020. Epub 2016/12/25. https://doi.org/10.1182/blood-2016-05-714089. PubMed PMID: 28011678; PMCID: PMC5324714
69. Carter BZ, Mak PY, Mak DH, Ruvolo VR, Schober W, McQueen T, Cortes J, Kantarjian HM, Champlin RE, Konopleva M, Andreeff M (2015) Synergistic effects of p53 activation via MDM2 inhibition in combination with inhibition of Bcl-2 or Bcr-Abl in CD34+ proliferating and quiescent chronic myeloid leukemia blast crisis cells. Oncotarget 6(31):30487–30499. Epub 2015/10/03. https://doi.org/10.18632/oncotarget.5890. PubMed PMID: 26431162; PMCID: PMC4741546
70. Li L, Wang L, Li L, Wang Z, Ho Y, McDonald T, Holyoake TL, Chen W, Bhatia R (2012) Activation of p53 by SIRT1 inhibition enhances elimination of CML leukemia stem cells in combination with imatinib. Cancer Cell 21(2):266–281. https://doi.org/10.1016/j.ccr.2011.12.020. PubMed PMID: 22340598; PMCID: PMC3285436
71. Abraham SA, Hopcroft LE, Carrick E, Drotar ME, Dunn K, Williamson AJ, Korfi K, Baquero P, Park LE, Scott MT, Pellicano F, Pierce A, Copland M, Nourse C, Grimmond SM, Vetrie D, Whetton AD, Holyoake TL (2016) Dual targeting of p53 and c-MYC selectively eliminates leukaemic stem cells. Nature 534(7607):341–346. Epub 2016/06/10. https://doi.org/10.1038/nature18288. PubMed PMID: 27281222; PMCID: PMC4913876
72. Zhang B, Strauss AC, Chu S, Li M, Ho Y, Shiang KD, Snyder DS, Huettner CS, Shultz L, Holyoake T, Bhatia R (2010) Effective targeting of quiescent chronic myelogenous leukemia stem cells by histone deacetylase inhibitors in combination with imatinib mesylate. Cancer Cell 17(5):427–442. Epub 2010/05/19. doi:S1535-6108(10)00102-9 [pii]10.1016/j.ccr.2010.03.011. PubMed PMID: 20478526; PMCID: PMC2873971
73. Scott MT, Korfi K, Saffrey P, Hopcroft LE, Kinstrie R, Pellicano F, Guenther C, Gallipoli P, Cruz M, Dunn K, Jorgensen HG, Cassels JE, Hamilton A, Crossan A, Sinclair A, Holyoake TL, Vetrie D (2016) Epigenetic reprogramming sensitizes CML stem cells to combined EZH2 and tyrosine kinase inhibition. jCancer Discov 6(11):1248–1257. Epub 2016/11/04. https://doi.org/10.1158/2159-8290.CD-16-0263. PubMed PMID: 27630125
74. Gallipoli P, Cook A, Rhodes S, Hopcroft L, Wheadon H, Whetton AD, Jorgensen HG, Bhatia R, Holyoake TL (2014) JAK2/STAT5 inhibition by nilotinib with ruxolitinib contributes to the elimination of CML CD34+ cells in vitro and in vivo. Blood 124(9):1492–1501. https://doi.org/10.1182/blood-2013-12-545640. PubMed PMID: 24957147; PMCID: PMC4148771
75. Neviani P, Harb JG, Oaks JJ, Santhanam R, Walker CJ, Ellis JJ, Ferenchak G, Dorrance AM, Paisie CA, Eiring AM, Ma Y, Mao HC, Zhang B, Wunderlich M, May PC, Sun C, Saddoughi SA, Bielawski J, Blum W, Klisovic RB, Solt JA, Byrd JC, Volinia S, Cortes J, Huettner CS, Koschmieder S, Holyoake TL, Devine

S, Caligiuri MA, Croce CM, Garzon R, Ogretmen B, Arlinghaus RB, Chen CS, Bittman R, Hokland P, Roy DC, Milojkovic D, Apperley J, Goldman JM, Reid A, Mulloy JC, Bhatia R, Marcucci G, Perrotti D (2013) PP2A-activating drugs selectively eradicate TKI-resistant chronic myeloid leukemic stem cells. J Clin Invest 123(10):4144–4157. Epub 2013/09/04. https://doi.org/10.1172/JCI68951. PubMed PMID: 23999433; PMCID: PMC3784537

76. Gallipoli P, Pellicano F, Morrison H, Laidlaw K, Allan EK, Bhatia R, Copland M, Jorgensen HG, Holyoake TL (2013) Autocrine TNF-alpha production supports CML stem and progenitor cell survival and enhances their proliferation. Blood 122(19):3335–3339. https://doi.org/10.1182/blood-2013-02-485607. PubMed PMID: 24041577; PMCID: 3953090
77. Zhang B, Chu S, Agarwal P, Campbell VL, Hopcroft L, Jorgensen HG, Lin A, Gaal K, Holyoake TL, Bhatia R (2016) Inhibition of interleukin-1 signaling enhances elimination of tyrosine kinase inhibitor treated CML stem cells. Blood. https://doi.org/10.1182/blood-2015-11-679928. PubMed PMID: 27621307
78. Jaras M, Johnels P, Hansen N, Agerstam H, Tsapogas P, Rissler M, Lassen C, Olofsson T, Bjerrum OW, Richter J, Fioretos T (2010) Isolation and killing of candidate chronic myeloid leukemia stem cells by antibody targeting of IL-1 receptor accessory protein. Proc Natl Acad Sci U S A 107(37):16280–16285. Epub 2010/09/02. doi:1004408107 [pii]10.1073/pnas.1004408107. PubMed PMID: 20805474; PMCID: PMC2941341
79. Irvine DA, Zhang B, Kinstrie R, Tarafdar A, Morrison H, Campbell VL, Moka HA, Ho Y, Nixon C, Manley PW, Wheadon H, Goodlad JR, Holyoake TL, Bhatia R, Copland M (2016) Deregulated hedgehog pathway signaling is inhibited by the smoothened antagonist LDE225 (Sonidegib) in chronic phase chronic myeloid leukaemia. Sci Rep 6:25476. https://doi.org/10.1038/srep25476. PubMed PMID: 27157927; PMCID: PMC4860619
80. Jin L, Hope KJ, Zhai Q, Smadja-Joffe F, Dick JE (2006) Targeting of CD44 eradicates human acute myeloid leukemic stem cells. Nat Med 12(10):1167–1174. Epub 2006/09/26. https://doi.org/10.1038/nm1483. PubMed PMID: 16998484
81. Carter BZ, Mak PY, Mu H, Zhou H, Mak DH, Schober W, Leverson JD, Zhang B, Bhatia R, Huang X, Cortes J, Kantarjian H, Konopleva M, Andreeff M (2016) Combined targeting of BCL-2 and BCR-ABL tyrosine kinase eradicates chronic myeloid leukemia stem cells. Sci Transl Med 8(355):355ra117. Epub 2016/09/09. https://doi.org/10.1126/scitranslmed.aag1180. PubMed PMID: 27605552; PMCID: PMC5111086

Minimal Residual Disease in Acute Myeloid Leukemia

7

Alexandra Gomez-Arteaga and Monica L. Guzman

Abstract

Monitoring measurable (minimal) residual disease (MRD) in acute myeloid leukemia (AML) has greatly increased our ability to assess chemosensitivity to treatment as well as the duration of treatment responses. There is strong evidence to support its prognostic value for long-term outcomes at different time points and across assays and targets. It's role as a surrogate endpoint to define risk-adapted strategies is still under evaluation. In this chapter, we will discuss the definition of MRD in AML, the potential contribution of leukemia stem cells (LSCs) to MRD and we will review all the current approaches to assess residual disease including the 2018 European Leukemia Network (ELN) working group recommendations for MRD standardization in AML. In addition, a summary of MRD studies associated to prognosis will be described.

Keyword

Minimal residual disease · MRD · Acute myeloid leukemia · Leukemia stem cells · Leukemia-associated immunophenotypes · FLT3 · Runx1 · DNMT3A · IDH1 · IDH2

7.1 Introduction

How do we define response to therapy for any type of cancer? For years, in AML, cytomorphology has been used to assess response to induction chemotherapy, defining less than 5% of residual blast and normalization of blood counts as a complete response (CR) [1]. However, despite the advances in the field, relapses are still the main risk for death [2]; 60–85% of patients younger than 60 years old achieve a CR but less than 30–40% are alive after 5 years [3–6].

For more than 30 years we have recognized that there are residual undetected cells at the time of morphologic CR that are not effectively eliminated by current regimens and thus the need additional post-remission therapy, otherwise most patients would relapse [7]. Therefore, it has become a priority to improve the ability to detect and measure residual disease in AML. For example, an average 70 Kg adult can have ~0.8×10^{12} nucleated cells in the bone marrow (BM), so even in morphologic CR, less than 5% of myeloblasts can account up to 10^{10} residual leukemic cells [8]. In the past years, there has been an incredible advance in available technologies to assess residual disease. Cytogenetics increased the ability to

A. Gomez-Arteaga · M. L. Guzman (✉)
Department of Medicine, Division of Hematology and Oncology, Weill Cornell Medicine – New York Presbyterian Hospital, New York, NY, USA
e-mail: alg9117@nyp.org; mlg2007@med.cornell.edu

J. A. Aguirre-Ghiso (ed.), *Biological Mechanisms of Minimal Residual Disease and Systemic Cancer*, Advances in Experimental Medicine and Biology 1100,
https://doi.org/10.1007/978-3-319-97746-1_7

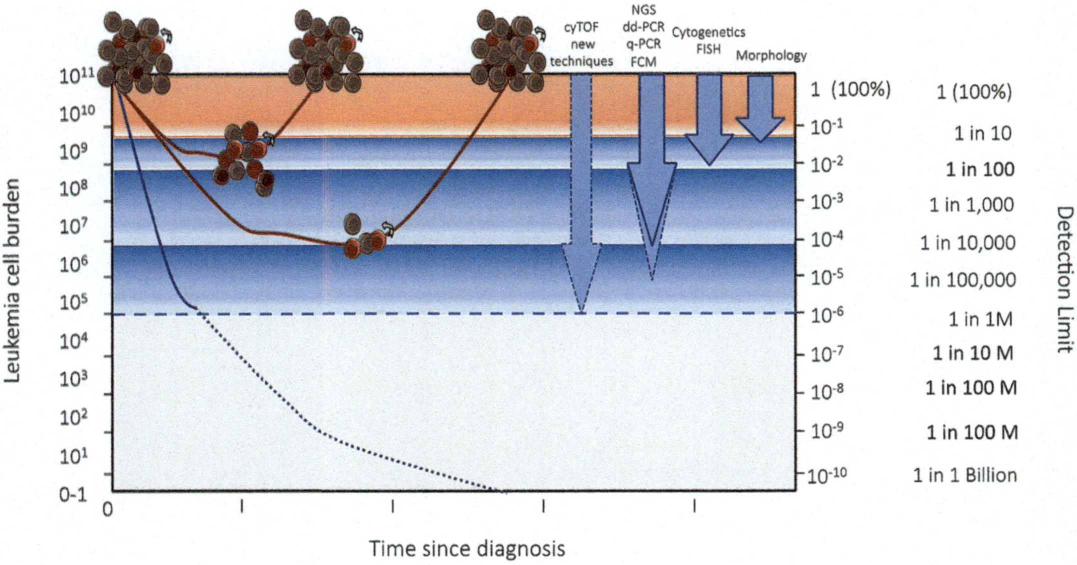

Fig. 7.1 Residual leukemia and likelihood of subsequent relapse

detect 1 leukemic cell in 100 (sensitivity of 10^{-2}), flow cytometry and PCR techniques can detect 1 in 1000 to 1 in 100,000 (10^{-3} to 10^{-5}), and newer approaches are reaching up to 1 in 1000,000 (10^{-6}) (Fig. 7.1).

7.2 Definition of Minimal Residual Disease

Minimal residual disease (MRD) is the terminology to define the number of leukemic cells that remain after the treatment. Because of the impact of MRD on disease progression, the ability to detect residual tumor burden must be at a high sensitivity. In theory, if the sensitivity increases to a maximum, then there will be higher chances of achieving a longer relapse free survival (RFS) and potentially achieve a cure.

A concern with the term "minimal", is that it is dependent on the sensitivity of the particular approach utilized to measure. Therefore, a more accurate terminology proposed is "measurable" thus MRD will reflect **measurable residual disease**.

A good MRD approach should be sensitive and highly specific. It should be a reproducible and a standardized procedure with ideally a wide interinstitutional validation. Current techniques can reach sensitivities between 10^{-3} and 10^{-6}.

There are two main clinical approaches to use MRD. The first one is to use it as a **prognostic marker** and the second one as **a surrogate end-point** to define response to treatment and to define treatment allocation. MRD is a time-point dependent variable. In AML, the prognostic significance of each time point is well reported (e.g. end of induction, end of consolidation, before transplant). The European Leukemia Net (ELN) proposed guidelines in 2017 included for the first time a definition for CR without MRD (CR_{MRD-}) as part of the response criteria [9]. However, unlike acute lymphoblastic leukemia (ALL), the treatment allocation based on MRD is not yet standardized and prospective adaptive trials are currently ongoing.

7.2.1 Leukemia Stem Cells (LSCs) and MRD

LSCs are a subpopulation of cells that possess stem cell properties such as cell renewal, quiescence and ability to generate progeny. In addition, LSCs have shown to be resistance to standard chemotherapy agents [10–12]. LSCs

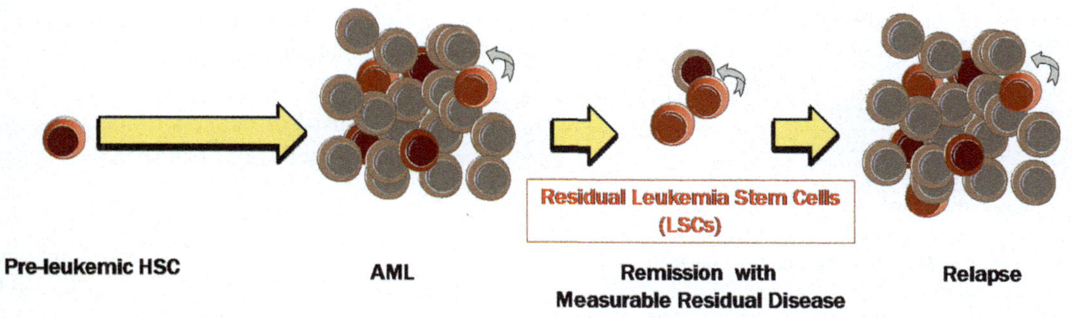

Fig. 7.2 Normal hematopoietic stem cells can acquire mutations that eventually lead to disease development. Current chemotherapeutics fail to eliminate all leukemic stem cells (LSCs). Residual LSCs provide a reservoir of malignant cells capable of initiating relapse

thus have the capacity to initiate disease and drive relapse (Fig. 7.2). LSCs can also be referred as leukemia-initiating cells (LIC) or leukemia propagating cells, as they were first demonstrated to be the AML fraction capable to initiate human leukemia in immunodeficient mice [13].

Next generation sequencing technologies has provided a tool to characterize the changes in the genomic composition of AML over time. A number of studies, isolating hematopoietic stem cells (HSCs) and putative LSCs by their specific immunophenotypic pattern, and using single cell sequencing, have identified that stem cells can harbor mutations in genes that are commonly involved in epigenetic regulation (i.e., DNMT3A, ASXL1, IDH2, and TET2) [14]. Such mutations can confer a competitive self-renewal advantage to HSCs, giving rise to pre-leukemic HSCs. They constitute the early steps in leukemogenesis and also play a key role in disease progression and clonal evolution. Therefore, it has been postulated that residual LSCs or pre-LSCs after therapy are the main drivers of relapse [15, 16] (Fig. 7.2).

In this chapter, we will discuss the state-of-the-art techniques that have been applied in the laboratory to measure MRD in AML, their prognostic significance and any evidence guiding the treatment strategy based on this knowledge. qPCR for PML-RAR for acute promyelocytic leukemia (APL) is widely standardized an in use, thus the main focus will be on non-APL AML.

7.3 Approaches to Assess MRD

7.3.1 Multi-parameter Flow Cytometry

Multi-parameter flow cytometry (MFC) has evolved as one of the most used technique to detect MRD in hematologic malignancies. It involves the specificity of the antibodies to detect aberrant cellular targets as well as the ability of the system to assess simultaneously multiple parameters in a single cell. The aim is to detect specific immunophenotypes that are characteristic to the cancer cell.

Designing and validating specific multicolor panels and standardizing the analysis, has been a priority for the past years in order to be able to compare and interpret the results across institutions. One of the groups that has implemented guidelines is the EuroFlow consortium which objective is to optimize and standardize instrument settings, antibody panels, sample preparation protocols, software for data analysis, and disease classification protocols [17–19]. Multicenter groups, such as the French Groupe d'Etude Immunologique des Leucémies (GEIL), have also adopted individual antibody panels and analysis recommendations [20].

7.3.1.1 Strategies for Using MFC to Detect MRD in AML

Leukemia-Associated Immunophenotypes (LAIP) By using screening antibody panels (see

Table 7.1 Immunophenotypic markers used to identify MRD by MFC

Panels	Markers
Panels to identify LAIP	
Pan leukocyte marker	CD 45
Primitive marker	CD34, CD117, CD123
Myeloid antigen	CD33, CD13
Aberrant marker expression: (these targets can also be used in **different-than-normal antigen expression**)	
Overexpression/under expression (20–30%)	dim CD33, HLA DR weak, overexpression of CD45, CD123, CD64
Cross-lineage expression (30–40%)	T-cell marker (CD2, CD3, CD5, CD7); B-cell marker (CD19); NKT marker (CD56)
Asynchronous expression (60–70%)	Myelomonocytic maturation (CD11b, CD15, CD4, CD64, CD36, CD 14)
Aberrant light scatter	Abnormal FSC/SSC
Panel targeting subpopulation of LSCs CD44, CLL-1, TIM-3, CD7, (CD34+/CD38−)	CD38, CD45RA, CD123, CD33, CD44, CLL-1, TIM-3, CD-7, CD11b, CD22, CD56)

Adapted from Refs. [10, 23, 24]

Table 7.1), it identifies antigen combinations that are characteristically expressed by the leukemic cells at diagnosis and not present in normal HSCs. This immunophenotypic pattern is known as leukemia-associated immunophenotype (LAIP). It is monitored through treatment, looking for residual leukemic cells expressing same aberrancy detected on the initial diagnosis sample. In order to include an antigen or a combination of antigens as a LAIP, it should be present in at least 5% to 10% of leukemic cells (to be sufficiently sensitive to monitor and detect at follow up), and sufficiently aberrant to be present on less than 0.1% to 0.01% of normal nucleated bone marrow (BM) cells (highly specific to avoid false positives) [19, 21]. There are more than 100 LAIPs that have been described, and about 90% of AMLs have a LAIP to follow [22, 23]. In an adequate sample, i.e. sufficient number of nucleated cells for analysis, a MRD detection sensitivity 1 in 1000 or 1 in 10,000 cells (10^{-3}–10^{-4}) can be achieved. Most MRD panels consist of 8–10 colors. A maximum sensitivity up to 10^{-5} has been described when at least 10×10^6 nucleated cells are stained for each MRD tube and there is systematic acquisition of $\geq 5 \times 10^6$ cells [24].

MFC panels consist of markers to identify cells in the myeloid/blast lineage; panels include a pan leukocyte marker (CD45), primitive markers (CD34, CD117, CD133), myeloid antigens (CD33, CD13) and 1 or more markers/marker combinations defined as aberrantly expressed. The assessment consists of the identification of antigen overexpression or underexpression (e.g. dim CD33, CD13, HLA DR weak), asynchronous maturation (e.g. myelomonocytic maturation expressing CD11b, CD14), cross-lineage antigen expression (e.g. lymphoid markers e.g. CD56, CD19, CD7, CD2), and/or aberrant light scatter to detect changes in cell size and granularity of the normal HSCs compartment (FSC/SSC) [24].

The prognostic impact of LAIP for MRD assessment has been demonstrated in several studies (Table 7.3) and it is, to date, the one most advanced despite limitations [25–29]. However, the disadvantage of LAIP-MFC, is that it can miss clonal evolution and if relapsing leukemic cells change antigen expression (referred as `immunophenotypic shift´), false negative MRD results can be reported [30]. In spite of the theoretical sensitivity, the individual sensitivity varies by LAIP used; therefore most studies using LAIPs for prognosis have chosen cutoffs of 0.01–0.1.

Different-from-Normal (DfN) It compares how the aberrant antigens from a specific cell population deviate from the antigenic-expression patterns seen in normal cells at specific stages of differentiation/maturation [31–33]. The main advantage of this approach is that it does not need prior knowledge of the specific immunophenotype of the leukemia cells at diagnosis and it can immunophenotypic shift. It uses set panels with combinations of the possible aberrantly expressed antigens at each time point of assessment [24].

DfN and LAIP are not freely interchangeable as it requires specific panels and methods for analysis. Most institutions use one of the two methodologies based on preference and resources.

Assessment of LSCs This approach measures the putative malignant hematopoietic stem cell subpopulations (identified by $CD34^+$/$CD38^-$ immunophenotype).

LSCs have been confirmed to be a prognostic marker, if there are higher frequencies of LSCs at diagnosis there is higher risk of remaining MRD positive after treatment and independently have a higher risk of relapse and decreased overall survival (OS) [34–36]. Residual detection of LSCs after induction, consolidation or prior to allo-SCT is also a poor prognostic factor [34, 37]. A recent report suggests that using the evaluation of the LSC population as an MRD method may be a more sensitive predictor of relapse after allogeneic transplant compared to DfN approach, making LSC assessment an attractive MRD approach to further study [38].

The prognostic impact of the LSC population has increased the interest of improving the techniques for monitoring LSCs, with the purpose to achieve the highest sensitivity to distinguish LSCs from normal HSC, which are also CD34+/CD38-. To this end, a new single 8-color detection tube has been described [11], it includes immunophenotypic markers that have been reported to distinguish LSCs from HSCs, such as CD45RA, CD123, CD44, CLL-1 and TIM-3 [10, 37].

LAIP-Based DfN Approach The recently released ELN guidelines for MRD monitoring in AML recommend that both LAIP and DfN approaches should be combined as the standardized method [39]. It will allow the detection of new aberrancies emerging at follow-up and monitor patients when there is an absence of diagnostic information. The adoption of this strategy in a multi -institutional level will depend on future validation studies (see below).

7.3.1.2 Advantages and Limitations of Using MFC

The advantages of MFC for MRD detection are that it is widely applicable, it is quantifiable, and it can estimate the number of aberrant cells. A fast turn-around is typically expected. It can collect other parameters such as cell viability. With the new consensus by the ELN working party that specifies the antibody panels, the number of parameters to evaluate per tube, the sample preparation, the cutoff used for positivity, and the number of events needed to achieve sensitivity it is expected that main disadvantages with FCM, the lack of standardization, will now be overcome. Furthermore, it is anticipated that this combined approach will be applicable to almost all patients. One of main disadvantages of MFC is the need of properly trained pathologist with adequate expertise to analyze the data [21, 39].

Currently, further multicenter validation of the LAIP-based DfN and LSC approaches are on their way. Such validations will allow for the determination of the prognostic significance of an MRD burden and the implementation of the appropriate cut-offs. In addition, the establishment of the optimal MRD monitoring intervals to define progression/impending relapse is also needed.

7.3.2 Quantitative Real-Time PCR (qPCR)

qPCR is commonly used to identify chimeric fusion genes (e.g. Philadelphia chromosome BCR-ABL, RUNX1-RUNX1T1 and CBFB-MYH11), gene rearrangements (e.g. TCR gene rearrangement), insertions/duplications (e.g. NPM1), point mutations (e.g. CEBPA, TP53, JAK2), and overexpressed genes (e.g. WT1).

The main considerations in order to have a reliable PCR method for detection of MRD are [40, 41]:

Analytical sensitivity: the minimum number of copies in a sample that can be detected with reasonable certainty (95% probability is commonly used) or *limit of detection* (LOD). This can vary by experimental procedures such as method of extraction and sample handling but speaks of the efficiency and robustness of the assay. Most qPCR techniques for MRD report a sensitivity of 1 in 10,000 which is the recommended by the Europe Against Cancer (EAC) program [41] but can vary between 10^{-2} to 10^{-6}.

Analytical specificity: Primers and probes used to detect the target of interest need to be highly specific to appropriate target sequence for genetic aberrations; if the specific primer is not included it can report a false negative. e.g. 90% of the mutations for NPM1 are type A, B, D, the other 10% need specific primers [42].

Reproducibility: The ELN working group recommends that assays should be performed under standardized conditions as proposed by the EAC program [41].

Reporting is also standardized as the mutant transcript levels relative to expression of the *ABL1* reference gene (housekeeping gene) with the use of plasmid standards to quantitate the transcripts [41]. For rare mutations, difference-in-cycle-thresholds (Ct) method can be used [43]. Sensitivity to which the assay was performed should be reported; specially it the transcript was not detected.

Table 7.2 Summary of the ELN MRD working party consensus and recommendation for molecular assessment: qPCR

Standardized targets	Presence of NPM1 mutations and the fusion genes RUNX1-RUNX1T1, and CBFB-MYH11
Technical requirements	Detailed recommendations by the Europe Against Cancer initiative (EAC)
Source of sample	PB and BM
Threshold for reporting	Should be reported at a sensitivity level of at least 1 in 1000
Time points for Molecular MRD assessment	At minimum at diagnosis, after 2 cycles of standard induction/consolidation chemotherapy and after the end of treatment
	For patients undergoing alloSCT, MRD should be assessed within 4 weeks prior to alloSCT
Definitions of molecular MRD assessment (for patients in CR)	Complete molecular remission (CRMRD-): two successive MRD negative samples within an interval of ≥4 weeks
	Molecular persistence at low copy numbers: <100–200 copies/10^4 ABL copies
	Molecular progression: an increase of MRD copy numbers ≥1 log10 between any two positive samples. Molecular relapse: an increase of the MRD level of ≥1 log10 between two positive samples in a patient who previously tested negative

7.3.2.1 Standardized Targets Used to Detect qPCR MRD in AML

As explained before, different gene alterations have been explored for MRD monitoring, but not all of them have achieved standardization. Some mutations involve a common underlying biochemical mechanism and physiological consequence for leukomogenesis (disease defining mutations) and some are mutations that interact and are influenced by coexisting mutation. Their selection depends on the stability of the mutation across time. qPCR for NPM1, RUNX1-RUNX1T1 and CBFB-MYH1 are currently the preferred MRD measurement when present as they are widely standardized [9, 39, 41]. Table 7.2 summarizes the recent recommendations from the ELN-MRD working group.

Core Binding Leukemia qPCR of chimeric fusion genes in RUNX1-RUNX1T1 (t(8;21)(q22;q22.1)) and CBFB-MYH11 (inv(16)(p13.1q22)) present in 7% and 5% of AML patients, are disease defining mutations, extremely stable between diagnosis and relapse, and allow MRD monitoring with high sensitivity (10^{-4}–10^{-6}) [44]. Even though it is present in only a minority of patients, it is standard of care in these patients to monitor MRD [9, 39].

NPM1 The nucleophosmin 1 gene (NPM1) gene is one of the most clinically relevant marker for molecular monitoring in cytogenetically normal (CN) karyotype AML (CN-AML). It is present in about 30% of AML patients and it is a disease defining mutation. NPM1 mutation are typically 4-nucleotide frameshift insertions in exon 12; 90% against Type A, B and D. However, the mutations are complexed and patients can be segregated in almost 150 subgroups [42]. Targeting as many of the potential mutations is key to apply to larger patient populations and reliable monitor disease. It is an informative prognostic marker for response to therapy and long-term outcomes [9, 42, 45, 46].

Other Targets **WT1** The Wilms tumor gene is a transcription factor that is overexpressed in granulocyte-monocyte precursors and down-regulated in differentiated myeloid cells. As it is overexpressed in 80–90% of AML, it was initially reported as a promising potential marker for MRD and as a prognostic marker for long-term outcomes after induction and response to transplant [47–51]. The ELN selected an assay spanning exons 1 and 2 based on superior performance compared to assays targeting exons 6, 7 or 8 [52]. This region is also less prone to mutations than other exons. Based on their testing, a sensitivity of 1 in 10^4 can be achieved and reporting has been defined as overexpressed above background (defined as 250 WT1 copies or 50 WT1 copies/10^4 ABL copies in BM and PB, respectively). The main limitation of this assay has been the poor selectivity as WT1 can be expressed in other cell populations and therefore its utility as a MRD marker was questioned. The ELN working group 2018 consensus, has recommended against its use for MRD unless there is no availability of any other method. If used, PB is recommended over BM [39].

FLT3-ITD, FLT3-TKD, MLL-PTD, mutations on **NRAS, KRAS, IDH1, IDH2,** and expression levels of **EVI1** as single markers of MRD are not recommended because of technical limitations. For example, the necessity for clone-specific primer/probes or instability of the marker with frequent losses or gains of mutations at relapse [39]. It is acknowledged that monitoring such genes may still have a prognostic significance when used in combination with a second MRD marker.

7.3.2.2 Advantages and Limitations of Using qPCR

The main advantage is its widely standardization for the described targets. Any laboratory with standardized qPCR assays is able to implement them.

The main disadvantages include its limitation to specific genetically defined patients (about 40% of patients). Primers and plasmids for quantification need to be mutation-specific and are not often available for all possible mutations. As discussed for FCM, relapse or progression may occur from new clones that are missed by qPCR. Finally, inadequate sample storage, preparation, and nucleic acid quality, can yield highly variable results [40, 53].

7.3.3 Digital Droplet PCR (ddPCR)

ddPCR involves performing individual PCRs in a large number of separate reaction chambers (or reaction partitions). Millions of these partitions are created with dilutions of template such that at least some of them contain no copies of the target sequence(s) of interest. Results are obtained by counting the number of partitions in which the amplified target sequence is detected and the number of partitions in which there is no amplification, which converts the output from an analog signal to a digital signal [53]. The quantification of the mean number of target sequences per partition is done by applying a Poisson correction to the fraction of the positive partitions and compensating for the fact that more than one copy of template may be present in some partitions [53]. The technique allows for multiplexing, where it can detect multiple targets simultaneously. For example, Mencia-Trinchant et al. recently reported on the use of ddPCR to monitor NPM1. They developed a new multiplex assay to detect common and rare subtypes of mutated NPM1 without the need for individual primers [54].

Besides NPM1, other ddPCR assays that have been reported include DNMT3A, IDH1, IDH2,

FLT3, KIT, SRF2, PTPN11, U2AF1, NRAS, ASXL, SF3B1, TP53, RUNX1, however large studies evaluating its prognostic capabilities are still under development [54–56].

7.3.3.1 Advantages and Limitations of Using ddPCR

The main advantages are that it allows for the measurement of rare variants with higher precision (able to compute confidence interval), which translates in a high MRD sensitivity: 1 in 100,000 (10^{-5}). The molecular counting is more reproducible as it is done solely on counting the number of amplification reactions, "absolute quantification", which is different than qPCR that has multiple variable that affect the assay efficiency such as the need of calibration curves with plasmids. Finally, as reported, it allows multiplexing and this technique is expected to grow within the next few years.

The main limitation for ddPCR is it cost. In addition, similar to qPCR, specific assays needs to be developed. It is affected by sequence damage, assay inhibition, molecular dropout (when partitions containing 2 assays amplify only one target). The absolute precision depends on the mean number of molecules per partition (dictated by the original sample concentration and how it has been prepared)*(42)* [53].

7.3.4 Next Generation Sequencing (NGS)

Whole exome sequencing identified the landscape of the recurrent genetic mutations in AML and led to the new classification for AML with new risk-defined groups [57–59]. Now **targeted NGS panels** that allow for the assessment of the specific disease-related gene mutations in a single assay is more accessible. Using targeted NGS panels to monitor the clearance of mutations as a technique for MRD monitoring is currently being tested [60–62]. The two-targeted NGS techniques currently in use for MRD measurement are amplicon-based enrichment and hybridization with the use of error-correction to allow for higher sensitive [63–65].

Initial comparison studies assessing the performance of NGS to qPCR for specific mutations including FLT3-ITD and NPM1 found them to be concordant (up to 95%). However, they underscored the limitations to assess MRD by FLT3-qPCR as it required specific primers for each mutation and there was antigenic shift upon relapse. On the other hand, NGS could detect clonal evolution making it a more informative approach for MRD assessment [66]. Most recently DfN-MFC was compared to a 28-gene amplicon captured NGS. Both approaches were found to be concordant in 71% of the cases and they were also complementary, as their combination increased the predictive value. A variant allele frequency (VAF) <5% was used for clearance [61]. Most recently, a large prospective study, the HOVON trial, evaluated the use of targeted NGS for MRD monitoring [60]. 482 patients were tested by a 54-gene targeted approach; 89.2% had somatic driver mutations present at diagnosis and 51.4% has persisting mutations at the time of CR at highly variable variant allele frequencies (VAF 0.0002-0.47). The predominant persisting mutation were in DNMT3A (78.7%), TET2 (54.2%) and ASXL1 (51.6%) and these were considered to be part of clonal hematopoiesis (and possibly remaining in preleukemic clones). The presence of persistent mutations after induction (excluding those associated to clonal hematopoiesis) was associated to an increased cumulative incidence of relapse (5-years CIR 58.3% vs 33.9% ($p < 0.001$)).

NGS appears to be a promising approach that may be able to detect clonal evolution and more adequately monitor response to therapy. However, measurements of MRD using NGS techniques are still under development, and are not ready for routine application outside of clinical trials [39].

Applying NGS for MRD monitoring is limited by depth of sequencing and algorithms used for mutation calling in order to reliable detect persisting mutations above background error rates. Among the areas under current evaluation is the sensitivity and coverage that can be achieved for each mutation. The variant allele frequency (VAF) that would be most informative to determine clinically relevant residual disease.

The kinetics of clearance of individual mutations, time points of analysis, bioinformatics tools for analysis, and mathematical models that can predict outcomes individualized for patients. It is not clear whether purification of cell populations, such as the LSCs compartment, will provide more in-depth information to identify clones that are likely to contribute to disease relapse.

Persistence of mutations has been shown to be of prognostic significance, however mutations that are likely to be present in pre-leukemic clones (e.g. DNMT3A, TET2 and ASXL1) are being excluded from the MRD definition [60]. Such mutations have also been noted in healthy individuals with increasing frequency as they age. This phenomena is referred to as age-related clonal hematopoiesis (ARCH) or clonal hematopoiesis of indeterminate potential (CHIP) [67]. More recent studies have demonstrated that the presence of clonal hematopoiesis does increase the risk of an individual to progress to AML and is though that as additional mutations develop, these cells are transformed into leukemic blasts [68, 69].

It is important to note that germline mutations (e.g. RUNX1, GATA2, CEBPA, DDX41, ANKRD26) [70] should be excluded from MRD measurement, germline origin is often confirmed by DNA sequencing using skin biopsies, hair follicles or buccal swabs.

7.3.5 Single-Cell Mass Cytometry (cyTOF)

An emerging approach is cyTOF, which is a flow cytometry assay based on time-of-flight mass spectrometry using antibodies labeled with heavy-metal ions, permitting simultaneous assessment of large panels of markers [71] (i.e. more than 50 markers per cells have been reported for different applications in AML [72]). It can overcome some of the limitations of MFC including, difficulty in standardization of assays and overlap of signal between the fluorescent channels. It has the potential of lowering the limit of detection and increasing the sensitivity (if enough sample is available). Initial studies of its applications to detect LSCs and residual leukemia are undergoing [55, 73, 74]. Is prognostic role for MRD will need to be elucidated.

7.4 Prognostic Significance of MRD

Table 7.3 summarizes key studies evaluating the prognostic impact of MRD at different time points during treatment.

7.4.1 MRD as an Endpoint for Risk Adapted Therapy

To date, treatment intensification for MRD positive patients after induction is supported in ALL but no prospective study has evaluated this in AML compared to allo-SCT, a current standard of care. The following studies have described interventions based on retrospective studies that had an impact on outcomes and should be further studied on prospective trials before further recommendations. <u>They do not constitute current clinical practice</u>.

7.4.1.1 Risk Directed Therapy in CBF t(8,21) AML

Decision to transplant: For patients who do not achieve more than 3-log reduction in RUNX1/RUNX1T1 transcripts after second consolidation, it was suggested that allo-SCT reduced the cumulative incidence of relapse (CIR) and improve disease free survival (DFS) compared with chemotherapy (22.1% vs 78.9%; $p < .0001$ and 61.7% vs 19.6%; $p < .001$) respectively) [75].

Monitoring and interventions post-transplant: For patients who undergo allo-SCT based on high risk features (i.e. positive MRD or presence of c-kit), monitoring of RUNX1/RUNX1T1 transcripts after transplant can further stratify patients: <3-log reduction within 12 months and/or < 4-log reduction at ≥12 months was significantly related to outcomes. Patients who received a preemptive donor lymphocyte infusion (DLI) when the increase in RUNX1-RUNX1T1 transcripts was

Table 7.3 Selected studies evaluating MRD and prognostic significance

Method	Timing of MRD measurement	MRO cutoff	Outcomes associated to MRD negativity	No. of pts (applicability %)	Refs.
LAIP-MFC	1st CR after I_1/I_2/C	<0.1%	Improved CIR/OS	517 (69%)	Terwijn [26]
	1st CR after I_1/I_2/C	<0.1%	Improved CIR/OS	427 (93%)	Freeman [27]
	1st CR after two induction cycles	<0.035%	Improved CIR/OS	56	Venditti [28]
Different-than normal-MFC	Prior to AlloSCT	<0.1%	Improved CIR/PFS/OS	359 (100%)	Araki [31]
	Prior to AlloSCT	<0.1%	Improved CIR/PFS/OS	253 (100%)	Walter [32]
Both LAIP/different-than normal-MFC	Prior to AlloSCT (9 studies)	~0.1%	Improved CIR (HR 2.81), PFS (HR 2.41) and OS (HR 1.98)	Meta-analysis	Buckley [29]
LSCs-MFC	1st CR after I_1/I_2/C	0.0001–0.0003%	Improved OS	91	Terwijn [34]
	Prior to Allo-SCT	0.02%	Improved CIR/OS	101	Bradbury [38]
q-PCR RUNX1-RUNX1T1	After C2	≥3 log reduction	Improved CIR/OS	137	Zhu [75]
	In follow up (PB/BM)	<100 copies PB or 500 copies BM/10^5 ABL copies	Improved CIR/OS	163	Yin [44]
q-PCR CBFB/MYH11	In follow up or at the end of treatment (PB)	<10 copies/10^5 ABL copies	Improved CIR	115	
	End of treatment (BM)	<50 copies/10^5 ABL copies	Improved CIR/OS		
qPCR – NPM1	1st CR after treatment (PB)	<0.01%	Improved RFS/OS	194	Ivey [42]
	1st CR after treatment (BM)	Undetectable	Improved RFS/OS	129	Kronke [45]
	1st CR after I_1/I_2 (PB)	≥4 log reduction	Improved CIR/OS	131	Balsat [46]
qPCR – WTI	Prior to AlloSCT (4 studies)	50–70 copies of WT1/10^4 ABL1	Improved CIR (HR 9.53), PFS (HR 5.80) and OS (5.25)	Meta-analysis	Buckley [29]
NGS	1st CR after I	VAF 0.0002–0.47%	Clearance of non CH-mutations: improved CIR/OS	430	Jongen-Layrencic [60]
	Prior to Allo-SCT	VAF <5%	Improved CIR/OS	77	Getta [61]

≤1-log had significantly lower 1-year CIR than the patients that received it when the increase was >1-log (0 vs. 55.0%, P = 0.015) [76].

7.4.1.2 Risk Directed Therapy in NPM1 AML

DFS and OS was significantly improved by allo-SCT in patients with a < 4-log reduction in NPM1 peripheral blood MRD (HR 0.25) [46]. The ELN working group recommends consideration of salvage therapy if there is an upward trajectory of MRD (log increase in either BM or PB) [39].

7.4.1.3 Risk Directed Therapy After Transplant with Low Dose Azacitidine

The RELAZA2 trial, studied if preemptive low dose azacitidine could reduce the risk of relapse

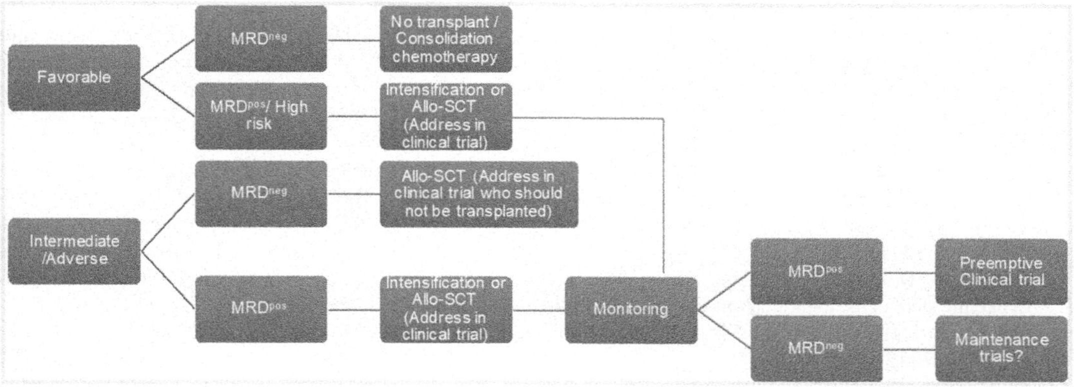

Fig. 7.3 Diagram for treatment allocation to clinical trial based on initial risk status and subsequent MRD surrogate endpoints. These recommendations are not standard

for patients who had received an allo-SCT and their chimerism analysis of CD34+ cells dropped to <80% [77]. This intervention stabilized or increased the chimerism in 80% of the 20 patients and delayed the relapse. Further studies of this intervention in patients who remained MRD positive after transplant will be needed.

7.5 Conclusion

The ability to assess MRD in a sensitive and objective manner is of critical importance for clinical practice to guide the care for patients with AML. There are still many unanswered questions that relate to risk-adapted decisions based on MRD definitions that will need to be further validated in prospective trials. With the new guidelines by the ELN working group it is expected that new multi-institutional collaborations will have a standardized framework to study intensification of therapy, drug approval, transplant eligibility, and maintenance strategies according to MRD status. Figure 7.3 has a diagram that shows our recommendations for treatment allocation to clinical trials based on sequential monitoring of MRD for treatment response. Adaptive trial designs are urgently needed.

References

1. Cheson BD, Bennett JM, Kopecky KJ, Buchner T, Willman CL, Estey EH et al (2003) Revised recommendations of the international working Group for Diagnosis, standardization of response criteria, treatment outcomes, and reporting standards for therapeutic trials in acute myeloid leukemia. J Clin Oncol 21(24):4642–4649. https://doi.org/10.1200/JCO.2003.04.036
2. Kantarjian H (2016) Acute myeloid leukemia--major progress over four decades and glimpses into the future. Am J Hematol 91(1):131–145. https://doi.org/10.1002/ajh.24246
3. Appelbaum FR (2012) Haematological cancer: the rule of three in AML induction--is cladribine the answer? Nat Rev Clin Oncol 19(9):376–377
4. Shah A, Andersson TM, Rachet B, Bjorkholm M, Lambert PC (2013) Survival and cure of acute myeloid leukaemia in England, 1971-2006: a population-based study. Br J Haematol 162(4):509–516. https://doi.org/10.1111/bjh.12425
5. Dohner H, Weisdorf DJ, Bloomfield CD (2015) Acute Myeloid Leukemia. N Engl J Med 373(12):1136–1152. https://doi.org/10.1056/NEJMra1406184
6. Koreth J, Schlenk R, Kopecky KJ, Honda S, Sierra J, Djulbegovic BJ et al (2009) Allogeneic stem cell transplantation for acute myeloid leukemia in first complete remission: systematic review and meta-analysis of prospective clinical trials. JAMA 301(22):2349–2361. https://doi.org/10.1001/jama.2009.813
7. Cassileth PA, Begg CB, Bennett JM, Bozdech M, Kahn SB, Weiler C et al (1984) A randomized study of the efficacy of consolidation therapy in adult acute nonlymphocytic leukemia. Blood 63(4):843–847

8. Hourigan CS, Gale RP, Gormley NJ, Ossenkoppele GJ, Walter RB (2017) Measurable residual disease testing in acute myeloid leukaemia. Leukemia 31(7):1482–1490. https://doi.org/10.1038/leu.2017.113
9. Dohner H, Estey E, Grimwade D, Amadori S, Appelbaum FR, Buchner T et al (2017) Diagnosis and management of AML in adults: 2017 ELN recommendations from an international expert panel. Blood 129(4):424–447. https://doi.org/10.1182/blood-2016-08-733196
10. Hanekamp D, Cloos J, Schuurhuis GJ (2017) Leukemic stem cells: identification and clinical application. Int J Hematol 105(5):549–557. https://doi.org/10.1007/s12185-017-2221-5
11. Zeijlemaker W, Kelder A, Oussoren-Brockhoff YJ, Scholten WJ, Snel AN, Veldhuizen D et al (2016) A simple one-tube assay for immunophenotypical quantification of leukemic stem cells in acute myeloid leukemia. Leukemia 30(2):439–446. https://doi.org/10.1038/leu.2015.252
12. Guzman ML, Allan JN (2014) Concise review: Leukemia stem cells in personalized medicine. Stem Cells (Dayton, Ohio) 32(4):844–851. https://doi.org/10.1002/stem.1597
13. Bonnet D, Dick JE (1997) Human acute myeloid leukemia is organized as a hierarchy that originates from a primitive hematopoietic cell. Nat Med 3(7):730–737
14. Jan M, Snyder TM, Corces-Zimmerman MR, Vyas P, Weissman IL, Quake SR et al (2012) Clonal evolution of preleukemic hematopoietic stem cells precedes human acute myeloid leukemia. Science Translational Medicine 4(149):149ra18. https://doi.org/10.1126/scitranslmed.3004315
15. Corces-Zimmerman MR, Hong WJ, Weissman IL, Medeiros BC, Majeti R (2014) Preleukemic mutations in human acute myeloid leukemia affect epigenetic regulators and persist in remission. Proc Natl Acad Sci U S A 111(7):2548–2553. https://doi.org/10.1073/pnas.1324297111
16. Sykes SM, Kokkaliaris KD, Milsom MD, Levine RL, Majeti R (2015) Clonal evolution of preleukemic hematopoietic stem cells in acute myeloid leukemia. Exp Hematol 43(12):989–992. https://doi.org/10.1016/j.exphem.2015.08.012
17. EuroFLow Consortium. ESLHO. https://euroflow.org/usr/pub/protocols.php
18. Kalina T, Flores-Montero J, van der Velden VH, Martin-Ayuso M, Bottcher S, Ritgen M et al (2012) EuroFlow standardization of flow cytometer instrument settings and immunophenotyping protocols. Leukemia 26(9):1986–2010. https://doi.org/10.1038/leu.2012.122
19. van Dongen JJ, Lhermitte L, Bottcher S, Almeida J, van der Velden VH, Flores-Montero J et al (2012) EuroFlow antibody panels for standardized n-dimensional flow cytometric immunophenotyping of normal, reactive and malignant leukocytes. Leukemia 26(9):1908–1975. https://doi.org/10.1038/leu.2012.120
20. Lacombe F, Campos L, Allou K, Arnoulet C, Delabarthe A, Dumezy F et al (2017) Prognostic value of multicenter flow cytometry harmonized assessment of minimal residual disease in acute myeloblastic leukemia. Hematol Oncol 36:422–428. https://doi.org/10.1002/hon.2488
21. Feller N, van der Velden VH, Brooimans RA, Boeckx N, Preijers F, Kelder A et al (2013) Defining consensus leukemia-associated immunophenotypes for detection of minimal residual disease in acute myeloid leukemia in a multicenter setting. Blood cancer journal 3:e129. https://doi.org/10.1038/bcj.2013.27
22. Del Principe MI, Buccisano F, Maurillo L, Sconocchia G, Cefalo M, Consalvo MI et al (2016) Minimal residual disease in acute myeloid leukemia of adults: determination, prognostic impact and clinical applications. Mediterr J Hematol Infect Dis 8(1):e2016052. https://doi.org/10.4084/MJHID.2016.052
23. Cruz NM, Mencia-Trinchant N, Hassane DC, Guzman ML (2017) Minimal residual disease in acute myelogenous leukemia. Int J Lab Hematol 39(1):53–60. https://doi.org/10.1111/ijlh.12670
24. Grimwade D, Freeman SD (2014) Defining minimal residual disease in acute myeloid leukemia: which platforms are ready for "prime time"? Blood 124(23):3345–3355. https://doi.org/10.1182/blood-2014-05-577593
25. Buccisano F, Maurillo L, Gattei V, Del Poeta G, Del Principe MI, Cox MC et al (2006) The kinetics of reduction of minimal residual disease impacts on duration of response and survival of patients with acute myeloid leukemia. Leukemia 20(10):1783–1789. https://doi.org/10.1038/sj.leu.2404313
26. Terwijn M, van Putten WL, Kelder A, van der Velden VH, Brooimans RA, Pabst T et al (2013) High prognostic impact of flow cytometric minimal residual disease detection in acute myeloid leukemia: data from the HOVON/SAKK AML 42A study. J Clin Oncol 31(31):3889–3897. https://doi.org/10.1200/JCO.2012.45.9628
27. Freeman SD, Virgo P, Couzens S, Grimwade D, Russell N, Hills RK et al (2013) Prognostic relevance of treatment response measured by flow cytometric residual disease detection in older patients with acute myeloid leukemia. J Clin Oncol 31(32):4123–4131. https://doi.org/10.1200/JCO.2013.49.1753
28. Venditti A, Buccisano F, Del Poeta G, Maurillo L, Tamburini A, Cox C et al (2000) Level of minimal residual disease after consolidation therapy predicts outcome in acute myeloid leukemia. Blood 96(12):3948–3952
29. Buckley SA, Wood BL, Othus M, Hourigan CS, Ustun C, Linden MA et al (2017) Minimal residual disease prior to allogeneic hematopoietic cell transplantation in acute myeloid leukemia: a meta-analysis. Haematologica 102(5):865–873. https://doi.org/10.3324/haematol.2016.159343
30. Cui W, Zhang D, Cunningham MT, Tilzer L (2014) Leukemia-associated aberrant immunophenotype

in patients with acute myeloid leukemia: changes at refractory disease or first relapse and clinicopathological findings. Int J Lab Hematol 36(6):636–649. https://doi.org/10.1111/ijlh.12193

31. Loken MR, Alonzo TA, Pardo L, Gerbing RB, Raimondi SC, Hirsch BA et al (2012) Residual disease detected by multidimensional flow cytometry signifies high relapse risk in patients with de novo acute myeloid leukemia: a report from Children's oncology group. Blood 120(8):1581–1588. https://doi.org/10.1182/blood-2012-02-408336
32. Araki D, Wood BL, Othus M, Radich JP, Halpern AB, Zhou Y et al (2016) Allogeneic hematopoietic cell transplantation for acute myeloid leukemia: time to move toward a minimal residual disease-based definition of complete remission? J Clin Oncol 34(4):329–336. https://doi.org/10.1200/JCO.2015.63.3826
33. Walter RB, Buckley SA, Pagel JM, Wood BL, Storer BE, Sandmaier BM et al (2013) Significance of minimal residual disease before myeloablative allogeneic hematopoietic cell transplantation for AML in first and second complete remission. Blood 122(10):1813–1821. https://doi.org/10.1182/blood-2013-06-506725
34. Terwijn M, Zeijlemaker W, Kelder A, Rutten AP, Snel AN, Scholten WJ et al (2014) Leukemic stem cell frequency: a strong biomarker for clinical outcome in acute myeloid leukemia. PLoS One 9(9):e107587. https://doi.org/10.1371/journal.pone.0107587
35. van Rhenen A, Feller N, Kelder A, Westra AH, Rombouts E, Zweegman S et al (2005) High stem cell frequency in acute myeloid leukemia at diagnosis predicts high minimal residual disease and poor survival. Clin Cancer Res 11(18):6520–6527. https://doi.org/10.1158/1078-0432.CCR-05-0468
36. Khan N, Freeman SD, Virgo P, Couzens S, Richardson P, Thomas I et al (2015) An immunophenotypic pretreatment predictor for poor response to induction chemotherapy in older acute myeloid leukaemia patients: blood frequency of CD34+ CD38 low blasts. Br J Haematol 170(1):80–84. https://doi.org/10.1111/bjh.13398
37. Gerber JM, Smith BD, Ngwang B, Zhang H, Vala MS, Morsberger L et al (2012) A clinically relevant population of leukemic CD34(+)CD38(−) cells in acute myeloid leukemia. Blood 119(15):3571–3577. https://doi.org/10.1182/blood-2011-06-364182
38. Bradbury C, Houlton AE, Akiki S, Gregg R, Rindl M, Khan J et al (2015) Prognostic value of monitoring a candidate immunophenotypic leukaemic stem/progenitor cell population in patients allografted for acute myeloid leukaemia. Leukemia 29(4):988–991. https://doi.org/10.1038/leu.2014.327
39. Schuurhuis GJ, Heuser M, Freeman S, Bene MC, Buccisano F, Cloos J et al (2018) Minimal/measurable residual disease in AML: consensus document from ELN MRD working party. Blood 131:1275–1291. https://doi.org/10.1182/blood-2017-09-801498
40. Bustin SA, Benes V, Garson JA, Hellemans J, Huggett J, Kubista M et al (2009) The MIQE guidelines: minimum information for publication of quantitative real-time PCR experiments. Clin Chem 55(4):611–622. https://doi.org/10.1373/clinchem.2008.112797
41. Gabert J, Beillard E, van der Velden VHJ, Bi W, Grimwade D, Pallisgaard N et al (2003) Standardization and quality control studies of 'real-time' quantitative reverse transcriptase polymerase chain reaction of fusion gene transcripts for residual disease detection in leukemia - a Europe against Cancer program. Leukemia 17(12):2318–2357. https://doi.org/10.1038/sj.leu.2403135
42. Ivey A, Hills RK, Simpson MA, Jovanovic JV, Gilkes A, Grech A et al (2016) Assessment of minimal residual disease in standard-risk AML. N Engl J Med 374(5):422–433. https://doi.org/10.1056/NEJMoa1507471
43. Grimwade D, Jovanovic JV, Hills RK, Nugent EA, Patel Y, Flora R et al (2009) Prospective minimal residual disease monitoring to predict relapse of acute promyelocytic leukemia and to direct pre-emptive arsenic trioxide therapy. J Clin Oncol 27(22):3650–3658. https://doi.org/10.1200/JCO.2008.20.1533
44. Yin JA, O'Brien MA, Hills RK, Daly SB, Wheatley K, Burnett AK (2012) Minimal residual disease monitoring by quantitative RT-PCR in core binding factor AML allows risk stratification and predicts relapse: results of the United Kingdom MRC AML-15 trial. Blood 120(14):2826–2835. https://doi.org/10.1182/blood-2012-06-435669
45. Kronke J, Schlenk RF, Jensen KO, Tschurtz F, Corbacioglu A, Gaidzik VI et al (2011) Monitoring of minimal residual disease in NPM1-mutated acute myeloid leukemia: a study from the German-Austrian acute myeloid leukemia study group. J Clin Oncol 29(19):2709–2716. https://doi.org/10.1200/JCO.2011.35.0371
46. Balsat M, Renneville A, Thomas X, de Botton S, Caillot D, Marceau A et al (2017) Postinduction minimal residual disease predicts outcome and benefit from allogeneic stem cell transplantation in acute myeloid leukemia with NPM1 mutation: a study by the acute leukemia French association group. J Clin Oncol 35(2):185–193. https://doi.org/10.1200/JCO.2016.67.1875
47. Nomdedeu JF, Esquirol A, Carricondo M, Pratcorona M, Hoyos M, Garrido A et al (2018) Bone marrow WT1 levels in allogeneic hematopoietic stem cell transplantation for acute myelogenous leukemia and myelodysplasia: clinically relevant time points and 100 copies threshold value. Biol Blood Marrow Transplant J Am Soc Blood Marrow Transplant 24(1):55–63. https://doi.org/10.1016/j.bbmt.2017.09.001
48. Pozzi S, Geroldi S, Tedone E, Luchetti S, Grasso R, Colombo N et al (2013) Leukaemia relapse after allogeneic transplants for acute myeloid leukaemia: predictive role of WT1 expression. Br J Haematol 160(4):503–509. https://doi.org/10.1111/bjh.12181
49. Guolo F, Minetto P, Clavio M, Miglino M, Galaverna F, Raiola AM et al (2017) Combining flow cytometry and WT1 assessment improves the prognostic value of pre-transplant minimal residual disease in acute

myeloid leukemia. Haematologica 102(9):e348–ee51. https://doi.org/10.3324/haematol.2017.167254
50. Israyelyan A, Goldstein L, Tsai W, Aquino L, Forman SJ, Nakamura R et al (2015) Real-time assessment of relapse risk based on the WT1 marker in acute leukemia and myelodysplastic syndrome patients after hematopoietic cell transplantation. Bone Marrow Transplant 50(1):26–33. https://doi.org/10.1038/bmt.2014.209
51. Candoni A, De Marchi F, Zanini F, Zannier ME, Simeone E, Toffoletti E et al (2017) Predictive value of pretransplantation molecular minimal residual disease assessment by WT1 gene expression in FLT3-positive acute myeloid leukemia. Exp Hematol 49:25–33. https://doi.org/10.1016/j.exphem.2017.01.005
52. Cilloni D, Renneville A, Hermitte F, Hills RK, Daly S, Jovanovic JV et al (2009) Real-time quantitative polymerase chain reaction detection of minimal residual disease by standardized WT1 assay to enhance risk stratification in acute myeloid leukemia: a European LeukemiaNet study. J Clin Oncol 27(31):5195–5201. https://doi.org/10.1200/JCO.2009.22.4865
53. Huggett JF, Foy CA, Benes V, Emslie K, Garson JA, Haynes R et al (2013) The digital MIQE guidelines: minimum information for publication of quantitative digital PCR experiments. Clin Chem 59(6):892–902. https://doi.org/10.1373/clinchem.2013.206375
54. Mencia-Trinchant N, Hu Y, Alas MA, Ali F, Wouters BJ, Lee S et al (2017) Minimal residual disease monitoring of acute myeloid leukemia by massively multiplex digital PCR in patients with NPM1 mutations. J Mol Diagn: JMD 19(4):537–548. https://doi.org/10.1016/j.jmoldx.2017.03.005
55. Pollyea DA, Stevens BM, Winters A, Minhajuddin M, Gutman JA, Purev E et al (2017) Venetoclax (Ven) with Azacitidine (Aza) for Untreated Elderly Acute Myeloid Leukemia (AML) Patients (Pts) Unfit for Induction Chemotherapy: Single Center Clinical Experience and Mechanistic Insights from Correlative Studies. Blood 130(1):181
56. Brambati C, Galbiati S, Xue E, Toffalori C, Crucitti L, Greco R et al (2016) Droplet digital polymerase chain reaction for DNMT3A and IDH1/2 mutations to improve early detection of acute myeloid leukemia relapse after allogeneic hematopoietic stem cell transplantation. Haematologica 101(4):e157–e161. https://doi.org/10.3324/haematol.2015.135467
57. Papaemmanuil E, Gerstung M, Bullinger L, Gaidzik VI, Paschka P, Roberts ND et al (2016) Genomic classification and prognosis in acute myeloid leukemia. N Engl J Med 374(23):2209–2221. https://doi.org/10.1056/NEJMoa1516192
58. Grimwade D, Ivey A, Huntly BJ (2016) Molecular landscape of acute myeloid leukemia in younger adults and its clinical relevance. Blood 127(1):29–41. https://doi.org/10.1182/blood-2015-07-604496
59. Cancer Genome Atlas Research N, Ley TJ, Miller C, Ding L, Raphael BJ, Mungall AJ et al (2013) Genomic and epigenomic landscapes of adult de novo acute myeloid leukemia. N Engl J Med 368(22):2059–2074. https://doi.org/10.1056/NEJMoa1301689
60. Jongen-Lavrencic M, Grob T, Kavelaars FG, Al Hinai ASA, Zeilemaker A, Erpelinck-Verschueren CAJ et al (2017) Prospective molecular MRD detection By NGS: a powerful independent predictor for relapse and survival in adults with newly diagnosed AML. Blood 130(1):LBA-5-LBA
61. Getta BM, Devlin SM, Levine RL, Arcila ME, Mohanty AS, Zehir A et al (2017) Multicolor flow cytometry and multigene next-generation sequencing are complementary and highly predictive for relapse in acute myeloid leukemia after allogeneic transplantation. Biol Blood Marrow Transplant J Am Soc Blood Marrow Transplant 23(7):1064–1071. https://doi.org/10.1016/j.bbmt.2017.03.017
62. Kohlmann A, Nadarajah N, Alpermann T, Grossmann V, Schindela S, Dicker F et al (2014) Monitoring of residual disease by next-generation deep-sequencing of RUNX1 mutations can identify acute myeloid leukemia patients with resistant disease. Leukemia 28(1):129–137. https://doi.org/10.1038/leu.2013.239
63. Schmitt MW, Kennedy SR, Salk JJ, Fox EJ, Hiatt JB, Loeb LA (2012) Detection of ultra-rare mutations by next-generation sequencing. Proc Natl Acad Sci U S A 109(36):14508–14513. https://doi.org/10.1073/pnas.1208715109
64. Hiatt JB, Pritchard CC, Salipante SJ, O'Roak BJ, Shendure J (2013) Single molecule molecular inversion probes for targeted, high-accuracy detection of low-frequency variation. Genome Res 23(5):843–854. https://doi.org/10.1101/gr.147686.112
65. Young AL, Wong TN, Hughes AE, Heath SE, Ley TJ, Link DC et al (2015) Quantifying ultra-rare pre-leukemic clones via targeted error-corrected sequencing. Leukemia 29(7):1608–1611. https://doi.org/10.1038/leu.2015.17
66. Thol F, Kolking B, Damm F, Reinhardt K, Klusmann JH, Reinhardt D et al (2012) Next-generation sequencing for minimal residual disease monitoring in acute myeloid leukemia patients with FLT3-ITD or NPM1 mutations. Genes Chromosomes Cancer 51(7):689–695. https://doi.org/10.1002/gcc.21955
67. Steensma DP, Bejar R, Jaiswal S, Lindsley RC, Sekeres MA, Hasserjian RP et al (2015) Clonal hematopoiesis of indeterminate potential and its distinction from myelodysplastic syndromes. Blood 126(1):9–16. https://doi.org/10.1182/blood-2015-03-631747
68. Ding L, Ley TJ, Larson DE, Miller CA, Koboldt DC, Welch JS et al (2012) Clonal evolution in relapsed acute myeloid leukaemia revealed by whole-genome sequencing. Nature 481(7382):506–510. https://doi.org/10.1038/nature10738
69. Desai P, Mencia-Trinchant N, Oleksandr S, Simon MS, Lee S, Ritchie EK et al (2017) Acute Myeloid Leukemia (AML) patients demonstrate increased prevalence of AML-defining mutations in peripheral blood years prior to the development of overt leukemia: a case-control study. Blood 130(1):408

70. Porter CC (2016) Germ line mutations associated with leukemias. Hematology Am Soc Hematol Educ Program 2016(1):302–308. https://doi.org/10.1182/asheducation-2016.1.302
71. Ornatsky O, Bandura D, Baranov V, Nitz M, Winnik MA, Tanner S (2010) Highly multiparametric analysis by mass cytometry. J Immunol Methods 361(1–2):1–20. https://doi.org/10.1016/j.jim.2010.07.002
72. Behbehani GK, Samusik N, Bjornson ZB, Fantl WJ, Medeiros BC, Nolan GP (2015) Mass cytometric functional profiling of acute myeloid leukemia defines cell-cycle and Immunophenotypic properties that correlate with known responses to therapy. Cancer Discov 5(9):988–1003. https://doi.org/10.1158/2159-8290.CD-15-0298
73. Han L, Qiu P, Zeng Z, Jorgensen JL, Mak DH, Burks JK et al (2015) Single-cell mass cytometry reveals intracellular survival/proliferative signaling in FLT3-ITD-mutated AML stem/progenitor cells. Cytometry Part A J Int Soc Anal Cytol 87(4):346–356. https://doi.org/10.1002/cyto.a.22628
74. Zeng Z, Konopleva M, Andreeff M (2017) Single-cell mass cytometry of acute myeloid leukemia and leukemia stem/progenitor cells. Methods Mol Biol 1633:75–86. https://doi.org/10.1007/978-1-4939-7142-8_5
75. Zhu HH, Zhang XH, Qin YZ, Liu DH, Jiang H, Chen H et al (2013) MRD-directed risk stratification treatment may improve outcomes of t(8;21) AML in the first complete remission: results from the AML05 multicenter trial. Blood 121(20):4056–4062. https://doi.org/10.1182/blood-2012-11-468348
76. Qin YZ, Wang Y, Xu LP, Zhang XH, Chen H, Han W et al (2017) The dynamics of RUNX1-RUNX1T1 transcript levels after allogeneic hematopoietic stem cell transplantation predict relapse in patients with t(8;21) acute myeloid leukemia. J Hematol Oncol 10(1):44. https://doi.org/10.1186/s13045-017-0414-2
77. Platzbecker U, Wermke M, Radke J, Oelschlaegel U, Seltmann F, Kiani A et al (2012) Azacitidine for treatment of imminent relapse in MDS or AML patients after allogeneic HSCT: results of the RELAZA trial. Leukemia 26(3):381–389. https://doi.org/10.1038/leu.2011.234

8 Characteristics and Therapeutic Targeting of Minimal Residual Disease in Childhood Acute Lymphoblastic Leukemia

Irmela Jeremias and Denis M. Schewe

Abstract

Acute lymphoblastic leukemia (ALL) is the most common childhood cancer. Early response to therapy, especially the measurement of minimal residual disease (MRD), remains the most reliable and strongest independent prognostic parameter. Intriguingly, little is known on the mechanisms sustaining MRD in that disease. Here, we summarize existing evidence on the influences of molecular genetics and clonal architecture of childhood ALL on disease persistence. Also, the impact of the leukemic niche on residual leukemia cells in the bone marrow and extramedullary compartments is reviewed. We further discuss existing *in vivo* models of minimal residual disease based on different cellular labelling strategies and engraftment of ALL cells in immunodeficient mouse strains. We finally draw some conclusions on potential strategies targeting residual ALL cells, with a focus on cellular and antibody-based immunotherapy.

Keyword

Acute lymphoblastic leukemia · Childhood leukemia · B-cell precursor · MRD · Dormancy · Mesenchymal stem cells · Quiescent cells · Xenograft models · Immunodeficient mice

I. Jeremias
Research Unit Apoptosis in Hematopoietic Stem Cells, Helmholtz Center Munich, German Center for Environmental Health (HMGU), Munich, Germany

Department of Pediatrics, Dr. von Hauner Childrens Hospital, Ludwig Maximilians University, Munich, Germany

German Consortium for Translational Cancer Research (DKTK), Partnering Site Munich, Munich, Germany
e-mail: irmela.jeremias@helmholtz-muenchen.de

D. M. Schewe (✉)
Department of Pediatrics I, Pediatric Hematology/Oncology, ALL-BFM Study Group, University Hospital Schleswig-Holstein, Kiel, Germany
e-mail: Denis.Schewe@uksh.de

8.1 Molecular Genetics of Childhood ALL

Acute lymphoblastic leukemia (ALL) is the most common childhood cancer and accounts for 20–25% of all cases [1]. The disease is characterized by immature lymphoid cells that acquire consecutive genetic lesions.

The first lesion can occur before birth, suggesting that some patients may harbor preleukemic clones prone to second hits and leukemia development over years [2]. According to single nucleotide polymorphisms and genome

J. A. Aguirre-Ghiso (ed.), *Biological Mechanisms of Minimal Residual Disease and Systemic Cancer*, Advances in Experimental Medicine and Biology 1100,
https://doi.org/10.1007/978-3-319-97746-1_8

wide association studies, around 12% of children with ALL develop leukemia based on germ line mutations [3]. This frequency may rise in the future, once further genetic alterations predisposing to leukemia have been discovered. Predominant genetic predispositions include Down syndrome and germline variants in certain driver genes such as *PAX5* or *p53* that can lead to leukemia development [4]. Leukemia initiation by environmental factors or infectious pathogens is sometimes hypothesized and modelled in preclinical settings [5, 6] but cannot yet be confirmed on a larger scale.

Nowadays, the diagnosis of ALL is made by a combination of morphological examination, immunophenotyping, cytogenetics and molecular genetic techniques in leukemic blasts from the peripheral blood or the bone marrow. With this diagnostic standard, the type of ALL (B-cell precursor (BCP) or T-lineage) and the differentiation stage of the leukemic cells as well as recurrent chromosomal or molecular abnormalities can be detected, some of which are associated with drug resistance, leukemia persistence and disease relapse. The classical genetic lesions in BCP-ALL are the *ETV6/RUNX1* translocation in 20–25%, hyperdiploidy in 20–25%, rearrangements involving the *KMT2A (MLL)* gene in 5% and the *BCR/ABL1* and *TCF3/PBX1* translocations, each in 3–4% of BCP-ALL cases.

Especially infants with *KMT2A* rearrangements show a propensity for therapy failure and leukemia persistence and therefore have a particularly poor prognosis [7]. The mechanisms governing residual disease in that entity are insufficiently understood. On the biological level, *KMT2A* rearranged infant BCP-ALL seems to be fueled by a dynamic back-and-forth between proliferating and relatively quiescent subclones, which however cannot be confined to a specific subset of cells [8]. Patients with the frequent *ETV6/RUNX1* translocation have a favorable prognosis, however the percentage of patients with late relapses years after termination of therapy is the highest among all ALL subgroups [9], suggesting a long term persistence of leukemic cells below the detection limits of contemporary methods.

Rarer BCP-ALL subtypes include hypodiploid BCP-ALL encompassing near-haploid (24–31 chromosomes) and low-hypodiploid (32–39 chromosomes) cases. Near-haploid BCP-ALL is characterized by alterations in the transcription factor *IKZF3* while low-hypodiploid BCP-ALL frequently harbors *TP53*, *RB1* and *IKZF2* mutations [10]. In low-hypodiploid ALL, the *TP53* mutations are germline in roughly half of the patients [10]. Both entities can have activated RAS and PI3K signaling and are associated with a very poor prognosis [10]. *TCF3/HLF* positive BCP-ALL occurs in <1% of patients and is almost universally fatal [11]. Many of the hypodiploid and *TCF3/HLF* positive patients are characterized by a failure of induction chemotherapy and thus, persistent residual disease [11–13]. *TCF3/HLF* positive patients showed a marked sensitivity to the Bcl-2 inhibitor venetoclax preclinically, providing a novel therapeutic option for the eradication of residual leukemia in that entity [11]. Similarly, a novel subgroup of patients with *MEF2D-BCL9* fusion show marked therapy resistance and disease persistence with a high *in vitro* sensitivity to molecular targeted agents such as bortezomib and vorinostat [14]. 10–15% of children with BCP-ALL suffer from a disease termed "*BCR/ABL*-like" or "Philadelphia-like" ALL, which has a gene expression profile similar to *BCR/ABL1* positive ALL but lacks the actual *BCR/ABL1* fusion [15, 16]. This disease entity is characterized by chromosomal translocations resulting in activated kinases as well as deletions in B-cell development genes such as *IKZF1*, *TCF3*, or *PAX5* [17, 18]. The therapeutic targeting of *ABL1* class fusions as well as *JAK2* and *CRLF2* abnormalities in "*BCR/ABL1* like" patients is a matter of ongoing investigation with the goal of eradicating residual leukemia by tyrosine kinase inhibitors. Recently, the IKZF1^{plus} signature encompassing deletions in *IKZF1*, known to confer a poor patient prognosis, co-occurring with deletions in *CDKN2A*, *CDKN2B*, *PAX5*, or *PAR1* in the absence of deletions in the ETS transcription factor *ERG* has been described as a very poor prognostic profile occurring in 6% of children with BCP-ALL [19]. It is important to note that the negative prognostic impact of IKZF1^{plus} was almost exclusively apparent when residual disease (MRD) was present [19], which suggests that essential cues on the mechanisms of MRD are concealed within that signature. A potential candidate for further studies may be the novel target *CTNND1*, encoding P120-catenin, which has been associated with treatment resistance in *IKZF1* deleted BCP-ALL in murine models [20, 21]. On the other hand, molecular events such as *DUX4* rearrangement accompanied by a transcriptional deregulation of *ERG* have been shown to be protective of the negative impact of *IKZF1* deletion [22]. Overexpression of CRLF2 can be considered a further hallmark of BCP-ALL subgroups with a high relapse risk, namely patients with *IGH-CRFL2* translocations and patients with *P2RY8-CRLF2* fusions [23]. It has however been suggested that *P2RY8-CRLF2* is lost at relapse most of the time in favour of *IKZF1* deletions promoting disease recurrence [24].

Recurrent molecular abnormalities in T-ALL mainly include the dysregulation of specific transcription factors, such as *TAL1, TAL2, TLX1, TLX3, HOXA, LMO1/LMO2, LMO2/LYL1* and *NKX2–1*, including a number of fusions with for example *MLLT10, KMT2A, ABL1, NUP98* and many other genes involving signaling pathways. Mutations in T-ALL are common and include *NOTCH1* mutations with a very high frequency. Early T cell precursor ALL (ETP-ALL), defined by a distinct profile including the aberrant expression of myeloid and stem cell markers on the cell surface has been associated with therapy resistance and disease persistence and has the worst prognosis among all T-ALL patients. ETP-ALL commonly showed deregulation of *LMO2/LYL1* and mutations of genes in the JAK-STAT or Ras signaling pathways, such as *FLT3, NRAS* and *JAK3* [25], leaving room for the development of novel therapeutic approaches. Although T-ALL has now been well characterized on the genomic level, recurrent alterations have not been included for risk stratification and therapy adaptation in most treatment protocols yet.

8.2 Characteristics of Cells in Minimal Residual Disease

Many clinical and biological factors predict patient outcome. Less favorable prognostic factors include male gender, higher initial white blood cell counts (>50.000/μl), infancy or adolescent age (>15 years), presence of central nervous system (CNS) involvement, a pro-B cell immune phenotype and some of the genetic factors described above. The intensity of therapy is adjusted based on these factors in many treatment protocols.

Childhood leukemia is a model disease demonstrating that treatment response represents the best prognostic parameter during anti-cancer treatment. In children with ALL, highly sensitive molecular techniques to quantify the tumor load persisting after treatment have been invented in the late 1980^{s}. This residual tumor burden has been termed residual or minimal residual disease (MRD) [26, 27]. MRD measurement was soon used to guide therapy intensity and to justify bone marrow transplantation due to its prognostic impact [28]. It then became clear that early treatment response, e. g. during and at the end of induction chemotherapy is highly predictive for overall outcome; as a consequence, early MRD measurement represents the most reliable and strongest independent prognostic parameter today [29, 30].

MRD is measured by multi-parametric flow cytometry or by PCR for patient specific fusion gene transcripts or immunoglobulin/T cell receptor (IG/TR) gene rearrangements at the end of induction therapy and during the first months thereafter. It is important to note that MRD describes leukemia persistence at low levels which cannot be detected by conventional cytomorphological methods, but that true MRD negativity cannot be defined [31]. MRD negativity by conventional assays may indicate the absence of leukemic cells in the patient analyzed, but may also be falsely influenced by a low number of cells subjected to the assay, the assay method itself (fusion gene versus IG/TR), direct effects of the therapy administered and the sensitivity limits of the PCR method [31]. Detectable MRD is clearly associated with a poor prognosis in upfront therapy of childhood ALL [29, 30].

The importance of MRD results from the fact that rising MRD levels almost always predict an impending relapse in both pediatric and adult patients, e. g. after allogeneic stem cell transplantation [32] or following immunotherapy [33]. As relapse is associated with an inferior and in some cases dismal prognosis, it is attractive to aim at preventing relapse development by eliminating the low amount of MRD cells present in a patient. It is therefore of utmost importance to learn about the mechanisms of MRD, which may significantly differ between different leukemic niches.

In the bone marrow, residual leukemic cells may display a phenotype of long term dormancy including properties of stemness and chemotherapy resistance [34]. Importantly, residual disease in the bone marrow seems to be highly influenced by the *in vivo* microenvironment and the removal of such dormant ALL cells from their niche can lead to a reversal of the phenotype by resumption of growth and reversal of chemotherapy resistance [34]. Models of the leukemic niche are still highly imprecise and it has been suggested that dormant leukemic cells rest in the endosteal niche in proximity to osteoblasts, similar to hematopoietic stem cells [35]. Osteopontin, an extracellular matrix molecule secreted by osteoblasts, may be important for the induction of cell cycle exit in these cells *in vivo* [36]. However, close communication of ALL cells with mesenchymal stem cells (MSCs) has also been found to play a role in disease persistence [35] and it could also be shown that the plasticity of MSCs within the leukemic niche (transition from nestin positive to alpha-smooth muscle actin positive MSCs with concomitant fibrosis of the surrounding tissue) can be significantly modulated by resting leukemic cells and by the effects of chemotherapy – to the disadvantage of the patient [37].

Unlike in acute myeloid leukemia, putative leukemic stem cells (i. e. "resting cells", "quiescent cells", "dormant cells", "leukemia initiating cells") in ALL could never be confined to a specific cellular compartment defined by an expression profile of surface markers. Leukemia initiating capacity is rather a stochastic cellular property, which can be acquired and turned on and off in a heterogeneous manner in all cells within an ALL patient [34, 38, 39]. This mirrors that ALL is subclonal in nature [40] and characterized by a coexistence of a variety of subclones that all have the capability to propagate the disease, e. g. in serial transplantation experiments in immunodeficient xenograft mice [41]. These cellular subclones at initial diagnosis can however have fundamentally divergent genetic backgrounds [40] even though in general, the genetic complexity of ALL is fairly limited compared to other neoplasms. It is thus under the pressure of intensive chemotherapy that subclones with a marked therapy resistance eventually emerge and fuel the disease at relapse. These emerging clones can then be more clearly defined on the genetic level. In BCP-ALL patients with early relapses, the rapid acquisition of subclonal diversity under the pressure of therapy has been shown to correlate with the emergence of fitter therapy resistant clonal progeny and an inferior patient prognosis while a "slow-to-quasi-inert" evolution pattern was associated with potential leukemic dormancy and late events [42].

8.3 Putative Therapeutic Targets in MRD Cells

In late relapsing *TEL/AML1* positive patients, longitudinal tracking of MRD cells can identify quiescent leukemic cells [43], suggesting that long-term dormancy plays an important role in that disease entity. From a clinical perspective however, late BCP-ALL relapses with *TEL/AML1* have a favorable prognosis [9]. In contrast, BCP-ALL and T-ALL cells at early relapse show a high prevalence of activating Ras pathway mutations (*KRAS, NRAS, PTPN11*), opening the therapeutic option of MEK inhibition [44, 45]. BCP-ALL cells from patients with Down syndrome or ETP-ALLs display a hyperactivation of JAK signaling [46, 47] and may profit from inhibitors of the IL7 receptor/JAK/STAT pathway such as the JAK inhibitor ruxolitinib. Patients with "*BCR/ABL1*-like" ALL may benefit from tyrosine kinase inhibitors, it is however unclear which inhibitor performs best in which patient. When approaches such as MEK inhibition, JAK inhibition or other tyrosine kinase inhibitors are considered to tailor future ALL therapies it must be considered that mutations commonly thought to be the drivers of relapse may play dual roles. Such mutations may support chemotherapy resistance and leukemia progression on the one hand, and drug sensitivity and the eradication of leukemia on the other, depending on the evolutionary pressure exerted by chemotherapy or activating mutations in other signaling pathways. This has been shown for relapsed childhood ALL, in which *RAS* mutant populations, almost universally subclonal at initial diagnosis, were either enriched or substituted with wildtype populations during the course of clonal evolution to relapse [48]. Interestingly, *RAS* mutations could lead to therapy resistance and to treatment sensitivity as well, dependent on the particular drug that was used [48]. Similarly, in infants with *KMT2A*-rearranged infant BCP-ALL, high FLT3 expression levels can confer therapy resistance and a poor prognosis in some patients [49], as well as an inhibition of cell cycle progression in the leukemic cells, therapy sensitivity and a favorable patient prognosis in others [50]. These examples highlight that a lot of therapeutic modalities are waiting in line with the goal to tackle residual disease which, nevertheless, require robust preclinical and clinical testing.

8.4 MRD in the Central Nervous System

MRD also exists in other body compartments, e. g. in extramedullary sanctuaries such as the central nervous system (CNS). CNS involvement is present in less than 5% of children at the onset of ALL. It is diagnosed when >5 white blood cells/µl cerebrospinal fluid (CSF) are measured upon examination of the CSF and ALL cells are morphologically identifiable in cytospins (CNS3 status). Although mostly localized in the meninges, a CNS3 status also occurs when neurological symptoms occur due to leukemic infiltration of the brain. Measurement of MRD by flow cytometry or patient-specific PCR for IG/TR rearrangements in the cerebrospinal fluid (CSF) may aid in the identification of CNS positive patients as the conventional cytomorphological method suffers from a rather low sensitivity. MRD measurements in the CSF may further describe patients with impending CNS relapses, however technical issues such as cellular degradation in CSF samples and a lack of appropriate controls currently prevent its broad clinical use.

It is important to note that most patients with isolated CNS relapse harbor MRD in the bone marrow [51], suggesting that CNS relapse is fueled by cell clones that acquired a propensity to enter or to migrate to this compartment at relapse. Alternatively, the CNS niche serves as a sanctuary site inducing quiescence and protecting residual ALL cells from the toxicities of chemotherapy. Indeed, BCP-ALL cells acquire a dormant phenotype in murine models when residing in the CNS niche, which is characterized by cell cycle arrest and resistance against commonly used CNS targeted agents such as methotrexate [52].

Several studies have identified molecules that may be responsible for the migration of ALL cells to the CNS including neural/glial antigen 2 [53], vascular endothelial growth factor [54], the chemokine receptors CXCR4 and CCR7 [55–57] and an interleukin-15 (IL-15) dependent upregulation of CXCR3 and other homing molecules [58]. IL-15 upregulation in leukemic cells has additionally been shown to be prognostic for CNS infiltration in children with ALL [59]. Other studies suggest that certain distinct pathways may play a role in inducing leukemic survival, such as the lectin galectin-3 secreted in the CNS niche and interacting with the Mer tyrosine kinase in *TCF3/PBX1* positive BCP-ALL cells [52], activation of the pre-B cell receptor signaling component Zap-70 in BCP-ALL [57] and upregulation of the transcription factor *PBX1* [60, 61].

A last component that may play a role in permitting MRD in the CNS is the fact, that the CNS niche is a largely immune privileged site. It could for example be shown that natural killer (NK) cells, commonly activated by a high expression of IL-15 in leukemic blasts, are very potent in controlling ALL in the blood, bone marrow and spleen in murine models [62]. However, their absence or relative paucity in the CNS may give rise to the survival of ALL cells in that niche [62]. This finding has important implications for CNS prophylactic therapy and emerging immunotherapeutic strategies in the future.

8.5 Preclinical *in vivo* Models of MRD

Studying the mechanisms of MRD in ALL cells requires appropriate pre-clinical *in vivo* models, which are scarce. Xenotransplantation models using NOD.Cg-Prkdc^{scid}Il2rg^{tm1Wjl}/SzJ (NSG) mice deficient for B/T lymphocytes as well as NK cells have become indispensable in ALL research, as patient samples with a number of cytogenetic and molecular alterations readily engraft these animals and human disease can be modelled in many aspects. Mice with similar levels of immunodeficiency such as Rag2$^{-/-}$ knockout mice or NOD.Cg-Prkdc^{scid} (NOD/SCID) mice with an intact NK cell compartment [62, 63] can alternatively be used for such approaches. Published experimental setups mimicking human MRD in xengraft mice mainly include label retention assays with membrane dyes identifying quiescent non-cycling ALL populations [36, 64, 65]. A certainly more sophisticated approach is the generation of patient derived xenograft ALL cells expressing multiple transgenes for *in vivo* imaging and for magnetic and fluorochrome sorting. Expression of the transgenes is then followed by a labelling step with carboxyfluorescein diacetate succinimidyl ester (CFSE) for the assessment of ALL cell proliferation [34]. This approach is a highly laborious and expensive effort. Furthermore, it requires key technologies and expertise that are available in only few research laboratories. However, the generation of label retaining cells with membrane dyes alone is simple to perform and can be used to answer a wide range of questions related to the mechanisms of ALL residual disease in the leukemic microenvironment such as in certain specialized regions of the hematopoietic niche in the bone marrow or in extramedullary compartments.

An alternative to both labelling approaches is the xenotransplantation of ALL cell numbers at limiting dilutions or the creation of an experimental therapy-induced MRD situation by applying combination chemotherapy in leukemic xenograft mice. Infant BCP-ALL cells carrying rearrangements of the *KMT2A* gene behave extremely aggressively in NSG mice and it could be shown that the orthotopic (directly into the bone marrow) transplantation of numbers as low as 100 cells per mouse reliably results in leukemic engraftment of 100% of the animals injected [66]. The transplantation of low ALL cell numbers into one femoral bone of a mouse mimics an MRD level of 10^{-4} to 10^{-6} in humans, a situation that is frequently encountered in the treatment of ALL in children and that can thus be generated for the exploration of novel therapies in the pre-clinical setting [66]. NSG mice and similarly immunodeficient animals tolerate high doses of chemotherapy with only minor clinical side effects (with the exception of anthracyclines) and human ALL blasts are in general sensitive to chemotherapy *in vivo*. It is therefore possible to create experimental MRD by allowing full leukemic engraftment (e. g. defined by a certain percentage of human leukemic blasts in the peripheral blood of the mice) and then treating the animals with regimens resembling induction chemotherapy in humans such as a combination of vincristine, dexamethasone and an asparaginase preparation [50, 66]. A drop in leukemic burden and MRD kinetics can easily be followed by quantitative PCR from murine bone marrow DNA, if the patient specific IG/TR rearrangements or the sequence of a specific fusion gene are known from the human disease [51, 66].

It has to be mentioned that the engraftment of human ALL cells in mice is largely dependent on the degree of immunodeficiency of the individual model. Therefore, further optimization of immunodeficient mouse models is a continuous effort on many ends. Further depletion of humoral or cellular components of the immune system (e. g. macrophages [66]) may render a model more permissive for the engraftment of ALL cells, especially of rare subsets. Approaches to introduce and to humanize individual components of the immune system in mice may be equally important, as these models will become relevant to elucidate the efficacy of the immune system to control residual disease in ALL and to identify the correct effector mechanisms for novel targeted therapies.

8.6 Immunotherapy of MRD

Immunotherapy is at present the most promising novel modality for the therapy of MRD in ALL. It appears likely that a high leukemic burden neutralizes the efficacy of any immunotherapeutic approach in leukemia, be it an antibody based strategy, a vaccination, or a therapy with engineered T cells. It is even possible that the administration of T cell products in situations of high leukemic burdens may harm the patient as transferred T cell products were shown to induce the proliferation of leukemic stem cells and cancer progression in experimental models of chronic myeloid leukemia [67]. Furthermore, life-threatening adverse events such as cytokine release syndrome are clearly potentiated when immunotherapy is applied during phases of a high leukemia loads in BCP-ALL [68, 69].

Current immunotherapies in childhood ALL are almost exclusively directed against the CD19 antigen and include CD19 chimeric antigen receptor (CAR) T cells (e. g. the FDA-approved product tisagenlecleucel) [70] and the [CD19 × CD3] bispecific T cell engager (BiTE) molecule blinatumomab [71]. In addition, promising results in individual patients have been reported with naked Fc engineered CD19 antibodies, which have been optimized for an enhanced recruitment of immunological effector cells expressing activating Fcγ receptors. These antibodies are expected to further expand the CD19 directed repertoire of immunotherapies in the near future [72]. All of these approaches are most efficient in MRD situations and rely on the ability of the therapy to evoke immune responses in the individual patient and on the presence of the target antigen in MRD cells. Blinatumomab is dependent on T cells and naked CD19 antibodies trigger killing mechanisms such as antibody dependent cell-mediated cytotoxicity (ADCC) by T and NK cells or antibody dependent cellular phagocytosis (ADCP).

Identifying immune escape mechanisms of residual leukemia cells such as the upregulation of programmed cell death 1 ligand 1 (PD-L1, CD274) counteracting the beneficial effects of blinatumomab [73], or the upregulation of CD47 inhibiting the activation of myeloid effector cells [74–78] will be important to improve the efficacy of immunotherapies for a wide range of hematological and solid malignancies in the near future. Importantly, the CD47 immune checkpoint can function as a therapeutic target by itself and preclinical experimental data on CD47 targeting constructs exist in ALL [75]. Clearly, drawbacks arise when immunotherapies are associated with antigen escape and outgrowth of antigen negative leukemic subclones as well as lineage switch, which can for example occur in *KMT2A* rearranged leukemia [79] due to the early differentiation stage of the cells in that entity. In BCP-ALL, CD19 negative relapses are observed both during blinatumomab and CAR T cell treatment in 10 to 20% of BCP-ALL cases [68, 80–83], suggesting that alternative treatment approaches are urgently needed in these cases. Valid options include the CD20 antibody rituximab [84] or CD22 specific immunotherapies (e. g. the antibody-drug conjugate inotuzumab ozogamicin [85] or CD22 CAR T-cells [86]) as well as immunotherapies against myeloid targets, which are sometimes co-expressed.

A further possible concept is the combination of two antibodies of different specificities for simultaneous targeting of two antigens on one cell, which may in theory prevent antigen negative relapses. Antibodies with limited efficacy such as Rituximab [84] may be potentiated by altering the activation status of immune effector cells such as NK cells [87] or by combining the antibody with immunomodulatory agents such as lenalinomide to promote the activation status of lymphocytes and/or myeloid effector populations [88, 89].

An important limitation of many immunotherapies is the fact that the microenvironments in which they preferentially act are poorly defined. For instance, CNS manifestations of ALL are unsatisfactorily accessible for antibodies due to the boundaries of the blood brain barrier so that the CNS niche may remain a protected sanctuary for ALL cells under immunotherapy as well. In contrast, CD19 specific CAR T cells have been detected in the CNS of patients [90] so that CAR T cells are considered an immunotherapeutic treatment option in CNS manifestations of ALL and their systemic use is currently expanded to diffuse large B cell lymphomas originating in the CNS [91]. Whether the tumor loads of ALL and lymphoma correlate with the occurrence of encephalopathy and other neurological side effects of the therapy currently remains unknown. A further important impediment for immunotherapy of ALL is the fact that for T-ALL no immunotherapeutic options exist at all, which is mainly due to a lack of potential cellular target antigens.

All in all, targeting MRD using immunotherapeutic approaches currently remains a most striving and promising area of research with very immediate clinical applications. The eradication of MRD by immunotherapy may result in long lasting cure of children with ALL. However, the best approaches and combination therapies still remain to be elucidated.

8.7 Conclusions

Understanding the mechanisms of minimal residual disease (MRD) is a major obstacle in continuing the success story of ALL therapy in children and adolescents. Conventional genetic and modern genomic characterization of ALL has identified a number of patient subgroups, in which persistent residual disease plays a major role in the definition of treatment response and patient prognosis. All novel therapeutic principles in ALL such as the inhibition of anti-apoptotic molecules, proteasome inhibition, tyrosine kinase inhibitors, MEK and JAK inhibition as well as the broad armamentarium of immunotherapies have the common goal of eradicating residual disease in specific molecular ALL subgroups and patient populations. Measurement of MRD by flow cytometry, IG/TR PCR and next generation sequencing techniques face numerous challenges but the technical advances of the last years have rendered MRD and its kinetics a reliable assessment of therapy response and prognostication in individual patients. Exploring the mechanisms of MRD requires appropriate *in vitro* labelling tools and *in vivo* models of ALL, which are almost exclusively based on xenografting patient cells into mice with different levels of immunodeficiency. This approach has, in the last years, identified a number of potential cues on how ALL cells survive in a state of MRD, including niche dependent mechanisms in the bone marrow and extramedullary sanctuaries, induction of dormancy and leukemia initiating potential as well as clonal selection under the pressure of therapy. From an immediate therapeutic perspective, widening the arsenal of immunotherapeutic approaches by novel constructs, novel targets and concomitant activation of specific effector mechanisms plays a predominant role.

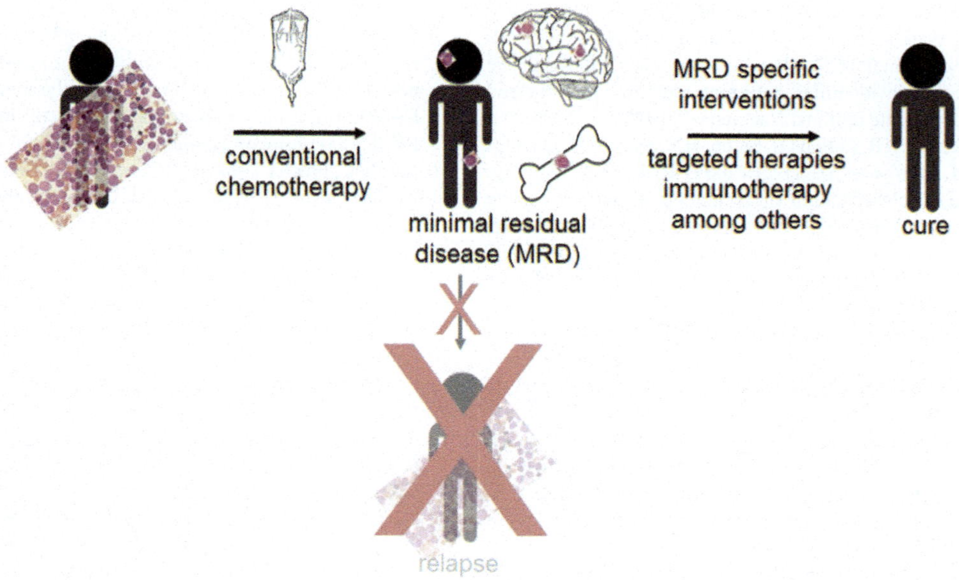

Eradicating MRD may be the optimal strategy to prevent relapse and provide permanent cure for children with acute lymphoblastic leukemia

Acknowledgement IJ was funded by ERC Consolidator Grant 681524; Mildred Scheel Professorship by German Cancer Aid; German Research Foundation (DFG), Collaborative Research Center 1243 "Genetic and epigenetic evolution of hematopoietic neoplasms", project A05; DFG proposal MA 1876/13-1; and by Bettina Bräu Stiftung and Dr. Helmut Legerlotz Stiftung. DS was funded by a Max Eder Fellowship by the German Cancer Aid, the Wilhelm-Sander-Stiftung and the Deutsche José-Carreras Leukämiestiftung.

References

1. Pui CH, Evans WE (2006) Treatment of acute lymphoblastic leukemia. N Engl J Med 354(2):166–178
2. Fischer S, Mann G, Konrad M, Metzler M, Ebetsberger G, Jones N et al (2007) Screening for leukemia- and clone-specific markers at birth in children with T-cell precursor ALL suggests a predominantly postnatal origin. Blood 110(8):3036–3038
3. Vijayakrishnan J, Kumar R, Henrion MY, Moorman AV, Rachakonda PS, Hosen I et al (2017) A genome-wide association study identifies risk loci for childhood acute lymphoblastic leukemia at 10q26.13 and 12q23.1. Leukemia 31(3):573–579
4. Hunger SP, Mullighan CG (2015) Acute lymphoblastic leukemia in children. N Engl J Med 373(16):1541–1552
5. Greaves M (2006) Infection, immune responses and the aetiology of childhood leukaemia. Nat Rev Cancer 6(3):193–203
6. Martin-Lorenzo A, Hauer J, Vicente-Duenas C, Auer F, Gonzalez-Herrero I, Garcia-Ramirez I et al (2015) Infection exposure is a causal factor in b-cell precursor acute lymphoblastic leukemia as a result of pax5-inherited susceptibility. Cancer Discov 5(12):1328–1343
7. Pieters R, Schrappe M, De Lorenzo P, Hann I, De Rossi G, Felice M et al (2007) A treatment protocol for infants younger than 1 year with acute lymphoblastic leukaemia (Interfant-99): an observational study and a multicentre randomised trial. Lancet 370(9583):240–250
8. Bardini M, Woll PS, Corral L, Luc S, Wittmann L, Ma Z et al (2015) Clonal variegation and dynamic competition of leukemia-initiating cells in infant acute lymphoblastic leukemia with MLL rearrangement. Leukemia 29(1):38–50
9. Forestier E, Heyman M, Andersen MK, Autio K, Blennow E, Borgstrom G et al (2008) Outcome of ETV6/RUNX1-positive childhood acute lymphoblastic leukaemia in the NOPHO-ALL-1992 protocol: frequent late relapses but good overall survival. Br J Haematol 140(6):665–672
10. Holmfeldt L, Wei L, Diaz-Flores E, Walsh M, Zhang J, Ding L et al (2013) The genomic landscape of hypodiploid acute lymphoblastic leukemia. Nat Genet 45(3):242–252

11. Fischer U, Forster M, Rinaldi A, Risch T, Sungalee S, Warnatz HJ et al (2015) Genomics and drug profiling of fatal TCF3-HLF-positive acute lymphoblastic leukemia identifies recurrent mutation patterns and therapeutic options. Nat Genet 47(9):1020–1029
12. Schultz KR, Pullen DJ, Sather HN, Shuster JJ, Devidas M, Borowitz MJ et al (2007) Risk- and response-based classification of childhood B-precursor acute lymphoblastic leukemia: a combined analysis of prognostic markers from the Pediatric Oncology Group (POG) and Children's Cancer Group (CCG). Blood 109(3):926–935
13. Schrappe M, Hunger SP, Pui CH, Saha V, Gaynon PS, Baruchel A et al (2012) Outcomes after induction failure in childhood acute lymphoblastic leukemia. N Engl J Med 366(15):1371–1381
14. Suzuki K, Okuno Y, Kawashima N, Muramatsu H, Okuno T, Wang X et al (2016) MEF2D-BCL9 fusion gene is associated with high-risk acute b-cell precursor lymphoblastic leukemia in adolescents. J Clin Oncol 34(28):3451–3459
15. Den Boer ML, van Slegtenhorst M, De Menezes RX, Cheok MH, Buijs-Gladdines JG, Peters ST et al (2009) A subtype of childhood acute lymphoblastic leukaemia with poor treatment outcome: a genome-wide classification study. Lancet Oncol 10(2):125–134
16. Mullighan CG, Su X, Zhang J, Radtke I, Phillips LA, Miller CB et al (2009) Deletion of IKZF1 and prognosis in acute lymphoblastic leukemia. N Engl J Med 360(5):470–480
17. Loh ML, Zhang J, Harvey RC, Roberts K, Payne-Turner D, Kang H et al (2013) Tyrosine kinome sequencing of pediatric acute lymphoblastic leukemia: a report from the Children's Oncology Group TARGET Project. Blood 121(3):485–488
18. Roberts KG, Li Y, Payne-Turner D, Harvey RC, Yang YL, Pei D et al (2014) Targetable kinase-activating lesions in Ph-like acute lymphoblastic leukemia. N Engl J Med 371(11):1005–1015
19. Stanulla M, Dagdan E, Zaliova M, Moricke A, Palmi C, Cazzaniga G et al (2018) IKZF1(plus) defines a new minimal residual disease-dependent very-poor prognostic profile in pediatric B-cell precursor acute lymphoblastic leukemia. J Clin Oncol 6(12):1240–1249 JCO2017743617
20. Witkowski MT, Hu Y, Roberts KG, Boer JM, McKenzie MD, Liu GJ et al (2017) Conserved IKAROS-regulated genes associated with B-progenitor acute lymphoblastic leukemia outcome. J Exp Med 214(3):773–791
21. Schjerven H, Ayongaba EF, Aghajanirefah A, McLaughlin J, Cheng D, Geng H et al (2017) Genetic analysis of Ikaros target genes and tumor suppressor function in BCR-ABL1(+) pre-B ALL. J Exp Med 214(3):793–814
22. Zhang J, McCastlain K, Yoshihara H, Xu B, Chang Y, Churchman ML et al (2016) Deregulation of DUX4 and ERG in acute lymphoblastic leukemia. Nat Genet 48(12):1481–1489
23. Cario G, Zimmermann M, Romey R, Gesk S, Vater I, Harbott J et al (2010) Presence of the P2RY8-CRLF2 rearrangement is associated with a poor prognosis in non-high-risk precursor B-cell acute lymphoblastic leukemia in children treated according to the ALL-BFM 2000 protocol. Blood 115(26):5393–5397
24. Vesely C, Frech C, Eckert C, Cario G, Mecklenbrauker A, Zur Stadt U et al (2017) Genomic and transcriptional landscape of P2RY8-CRLF2-positive childhood acute lymphoblastic leukemia. Leukemia 31(7):1491–1501
25. Liu Y, Easton J, Shao Y, Maciaszek J, Wang Z, Wilkinson MR et al (2017) The genomic landscape of pediatric and young adult T-lineage acute lymphoblastic leukemia. Nat Genet 49(8):1211–1218
26. Katz F, Ball L, Gibbons B, Chessells J (1989) The use of DNA probes to monitor minimal residual disease in childhood acute lymphoblastic leukaemia. Br J Haematol 73(2):173–180
27. Neale GA, Menarguez J, Kitchingman GR, Fitzgerald TJ, Koehler M, Mirro J Jr et al (1991) Detection of minimal residual disease in T-cell acute lymphoblastic leukemia using polymerase chain reaction predicts impending relapse. Blood 78(3):739–747
28. Brisco MJ, Condon J, Hughes E, Neoh SH, Sykes PJ, Seshadri R et al (1994) Outcome prediction in childhood acute lymphoblastic leukaemia by molecular quantification of residual disease at the end of induction. Lancet 343(8891):196–200
29. Conter V, Bartram CR, Valsecchi MG, Schrauder A, Panzer-Grumayer R, Moricke A et al (2010) Molecular response to treatment redefines all prognostic factors in children and adolescents with B-cell precursor acute lymphoblastic leukemia: results in 3184 patients of the AIEOP-BFM ALL 2000 study. Blood 115(16):3206–3214
30. Schrappe M, Valsecchi MG, Bartram CR, Schrauder A, Panzer-Grumayer R, Moricke A et al (2011) Late MRD response determines relapse risk overall and in subsets of childhood T-cell ALL: results of the AIEOP-BFM-ALL 2000 study. Blood 118(8):2077–2084
31. Bruggemann M, Kotrova M (2017) Minimal residual disease in adult ALL: technical aspects and implications for correct clinical interpretation. Blood advances. 1(25):2456–2466
32. Bader P, Kreyenberg H, von Stackelberg A, Eckert C, Salzmann-Manrique E, Meisel R et al (2015) Monitoring of minimal residual disease after allogeneic stem-cell transplantation in relapsed childhood acute lymphoblastic leukemia allows for the identification of impending relapse: results of the ALL-BFM-SCT 2003 trial. J Clin Oncol 33(11):1275–1284
33. Gokbuget N, Zugmaier G, Klinger M, Kufer P, Stelljes M, Viardot A et al (2017) Long-term relapse-free survival in a phase 2 study of blinatumomab for the treatment of patients with minimal residual disease in B-lineage acute lymphoblastic leukemia. Haematologica 102(4):e132–e1e5

34. Ebinger S, Ozdemir EZ, Ziegenhain C, Tiedt S, Castro Alves C, Grunert M et al (2016) Characterization of rare, dormant, and therapy-resistant cells in acute lymphoblastic leukemia. Cancer Cell 30(6):849–862
35. Pal D, Heidenreich O, Vormoor J (2016) Dormancy stems the tide of chemotherapy. Cancer Cell 30(6):825–826
36. Boyerinas B, Zafrir M, Yesilkanal AE, Price TT, Hyjek EM, Sipkins DA (2013) Adhesion to osteopontin in the bone marrow niche regulates lymphoblastic leukemia cell dormancy. Blood 121(24):4821–4831
37. Duan CW, Shi J, Chen J, Wang B, Yu YH, Qin X et al (2014) Leukemia propagating cells rebuild an evolving niche in response to therapy. Cancer Cell 25(6):778–793
38. le Viseur C, Hotfilder M, Bomken S, Wilson K, Rottgers S, Schrauder A et al (2008) In childhood acute lymphoblastic leukemia, blasts at different stages of immunophenotypic maturation have stem cell properties. Cancer Cell 14(1):47–58
39. Lang F, Wojcik B, Rieger MA (2015) Stem cell hierarchy and clonal evolution in acute lymphoblastic leukemia. Stem Cells Int 2015:137164
40. Anderson K, Lutz C, van Delft FW, Bateman CM, Guo Y, Colman SM et al (2011) Genetic variegation of clonal architecture and propagating cells in leukaemia. Nature 469(7330):356–361
41. Rehe K, Wilson K, Bomken S, Williamson D, Irving J, den Boer ML et al (2013) Acute B lymphoblastic leukaemia-propagating cells are present at high frequency in diverse lymphoblast populations. EMBO Mol Med 5(1):38–51
42. Spinella JF, Richer C, Cassart P, Ouimet M, Healy J, Sinnett D (2018) Mutational dynamics of early and late relapsed childhood ALL: rapid clonal expansion and long-term dormancy. Blood Adv 2(3):177–188
43. Lutz C, Woll PS, Hall G, Castor A, Dreau H, Cazzaniga G et al (2013) Quiescent leukaemic cells account for minimal residual disease in childhood lymphoblastic leukaemia. Leukemia 27(5):1204–1207
44. Irving J, Matheson E, Minto L, Blair H, Case M, Halsey C et al (2014) Ras pathway mutations are prevalent in relapsed childhood acute lymphoblastic leukemia and confer sensitivity to MEK inhibition. Blood 124(23):3420–3430
45. Richter-Pechanska P, Kunz JB, Hof J, Zimmermann M, Rausch T, Bandapalli OR et al (2017) Identification of a genetically defined ultra-high-risk group in relapsed pediatric T-lymphoblastic leukemia. Blood Cancer J 7(2):e523
46. Delgado-Martin C, Meyer LK, Huang BJ, Shimano KA, Zinter MS, Nguyen JV et al (2017) JAK/STAT pathway inhibition overcomes IL7-induced glucocorticoid resistance in a subset of human T-cell acute lymphoblastic leukemias. Leukemia 31(12):2568–2576
47. Schwartzman O, Savino AM, Gombert M, Palmi C, Cario G, Schrappe M et al (2017) Suppressors and activators of JAK-STAT signaling at diagnosis and relapse of acute lymphoblastic leukemia in Down syndrome. Proc Natl Acad Sci U S A 114(20):E4030–E40E9
48. Oshima K, Khiabanian H, da Silva-Almeida AC, Tzoneva G, Abate F, Ambesi-Impiombato A et al (2016) Mutational landscape, clonal evolution patterns, and role of RAS mutations in relapsed acute lymphoblastic leukemia. Proc Natl Acad Sci U S A 113(40):11306–11311
49. Stam RW, Schneider P, de Lorenzo P, Valsecchi MG, den Boer ML, Pieters R (2007) Prognostic significance of high-level FLT3 expression in MLL-rearranged infant acute lymphoblastic leukemia. Blood 110(7):2774–2775
50. Fedders H, Alsadeq A, Schmah J, Vogiatzi F, Zimmermann M, Moricke A et al (2017) The role of constitutive activation of FMS-related tyrosine kinase-3 and NRas/KRas mutational status in infants with KMT2A-rearranged acute lymphoblastic leukemia. Haematologica 102(11):e438–ee42
51. Alten J, Claviez A, Vieth S, Cario G, Schewe DM (2017) Mouse-MRD" in central nervous system acute lymphoblastic leukaemia: assessing bone marrow minimal residual disease using a xenograft model – from bedside to the bench and back again. Br J Haematol. https://doi.org/10.1111/bjh.15024
52. Krause S, Pfeiffer C, Strube S, Alsadeq A, Fedders H, Vokuhl C et al (2015) Mer tyrosine kinase promotes the survival of t(1;19)-positive acute lymphoblastic leukemia (ALL) in the central nervous system (CNS). Blood 125(5):820–830
53. Prieto C, Lopez-Millan B, Roca-Ho H, Stam RW, Romero-Moya D, Rodriguez-Baena FJ et al (2018) NG2 antigen is involved in leukemia invasiveness and central nervous system infiltration in MLL-rearranged infant B-ALL. Leukemia 32(3):633–644
54. Munch V, Trentin L, Herzig J, Demir S, Seyfried F, Kraus JM et al (2017) Central nervous system involvement in acute lymphoblastic leukemia is mediated by vascular endothelial growth factor. Blood 130(5):643
55. Jost TR, Borga C, Radaelli E, Romagnani A, Perruzza L, Omodho L et al (2016) Role of CXCR4-mediated bone marrow colonization in CNS infiltration by T-cell acute lymphoblastic leukemia. J Leukoc Biol 99:1077–1087
56. Buonamici S, Trimarchi T, Ruocco MG, Reavie L, Cathelin S, Mar BG et al (2009) CCR7 signalling as an essential regulator of CNS infiltration in T-cell leukaemia. Nature 459(7249):1000–1004
57. Alsadeq A, Fedders H, Vokuhl C, Belau NM, Zimmermann M, Wirbelauer T et al (2017) The role of ZAP70 kinase in acute lymphoblastic leukemia infiltration into the central nervous system. Haematologica 102(2):346–355
58. Williams MT, Yousafzai Y, Cox C, Blair A, Carmody R, Sai S et al (2014) Interleukin-15 enhances cellular proliferation and upregulates CNS homing molecules in pre-B acute lymphoblastic leukemia. Blood 123(20):3116–3127
59. Cario G, Izraeli S, Teichert A, Rhein P, Skokowa J, Moricke A et al (2007) High interleukin-15 expression characterizes childhood acute lymphoblastic leukemia with involvement of the CNS. J Clin Oncol 25(30):4813–4820

60. Gaynes JS, Jonart LM, Zamora EA, Naumann JA, Gossai NP, Gordon PM (2017) The central nervous system microenvironment influences the leukemia transcriptome and enhances leukemia chemo-resistance. Haematologica 102(4):e136–e1e9
61. Alsadeq A, Schewe DM (2017) Acute lymphoblastic leukemia of the central nervous system: on the role of PBX1. Haematologica 102(4):611–613
62. Frishman-Levy L, Shemesh A, Bar-Sinai A, Ma C, Ni Z, Frenkel S et al (2015) Central nervous system acute lymphoblastic leukemia: role of natural killer cells. Blood 125(22):3420–3431
63. Meyer LH, Debatin KM (2011) Diversity of human leukemia xenograft mouse models: implications for disease biology. Cancer Res 71(23):7141–7144
64. Colmone A, Amorim M, Pontier AL, Wang S, Jablonski E, Sipkins DA (2008) Leukemic cells create bone marrow niches that disrupt the behavior of normal hematopoietic progenitor cells. Science 322(5909):1861–1865
65. Sipkins DA, Wei X, Wu JW, Runnels JM, Cote D, Means TK et al (2005) In vivo imaging of specialized bone marrow endothelial microdomains for tumour engraftment. Nature 435(7044):969–973
66. Schewe DM, Alsadeq A, Sattler C, Lenk L, Vogiatzi F, Cario G et al (2017) An Fc engineered CD19 antibody eradicates MRD in patient-derived MLL-rearranged acute lymphoblastic leukemia xenografts. Blood 130:1620–1627
67. Schurch C, Riether C, Amrein MA, Ochsenbein AF (2013) Cytotoxic T cells induce proliferation of chronic myeloid leukemia stem cells by secreting interferon-gamma. J Exp Med 210(3):605–621
68. Maude SL, Frey N, Shaw PA, Aplenc R, Barrett DM, Bunin NJ et al (2014) Chimeric antigen receptor T cells for sustained remissions in leukemia. N Engl J Med 371(16):1507–1517
69. Porter D, Frey N, Wood PA, Weng Y, Grupp SA (2018) Grading of cytokine release syndrome associated with the CAR T cell therapy tisagenlecleucel. J Hematol Oncol 11(1):35
70. Maude SL, Laetsch TW, Buechner J, Rives S, Boyer M, Bittencourt H et al (2018) Tisagenlecleucel in children and young adults with B-cell lymphoblastic leukemia. N Engl J Med 378(5):439–448
71. von Stackelberg A, Locatelli F, Zugmaier G, Handgretinger R, Trippett TM, Rizzari C et al (2016) Phase I/phase II study of blinatumomab in pediatric patients with relapsed/refractory acute lymphoblastic leukemia. J Clin Oncol 34(36):4381–4389
72. Seidel UJ, Schlegel P, Grosse-Hovest L, Hofmann M, Aulwurm S, Pyz E et al (2016) Reduction of minimal residual disease in pediatric B-lineage acute lymphoblastic leukemia by an Fc-optimized CD19 Antibody. Mol Ther 24(9):1634–1643
73. Raedler H, Vieyra MB, Leisman S, Lakhani P, Kwan W, Yang M et al (2011) Anti-complement component C5 mAb synergizes with CTLA4Ig to inhibit alloreactive T cells and prolong cardiac allograft survival in mice. Am J Transplant Off J Am Soc Transplant Am Soc Transplant Surg 11(7):1397–1406
74. Weissman I (2016) How one thing led to another. Annu Rev Immunol 34:1–30
75. Chao MP, Alizadeh AA, Tang C, Jan M, Weissman-Tsukamoto R, Zhao F et al (2011) Therapeutic antibody targeting of CD47 eliminates human acute lymphoblastic leukemia. Cancer Res 71(4):1374–1384
76. Chao MP, Alizadeh AA, Tang C, Myklebust JH, Varghese B, Gill S et al (2010) Anti-CD47 antibody synergizes with rituximab to promote phagocytosis and eradicate non-Hodgkin lymphoma. Cell 142(5):699–713
77. Jaiswal S, Jamieson CH, Pang WW, Park CY, Chao MP, Majeti R et al (2009) CD47 is upregulated on circulating hematopoietic stem cells and leukemia cells to avoid phagocytosis. Cell 138(2):271–285
78. Ring NG, Herndler-Brandstetter D, Weiskopf K, Shan L, Volkmer JP, George BM et al (2017) Anti-SIRPalpha antibody immunotherapy enhances neutrophil and macrophage antitumor activity. Proc Natl Acad Sci U S A 114(49):E10578–E10585
79. Gardner R, Wu D, Cherian S, Fang M, Hanafi LA, Finney O et al (2016) Acquisition of a CD19-negative myeloid phenotype allows immune escape of MLL-rearranged B-ALL from CD19 CAR-T-cell therapy. Blood 127(20):2406–2410
80. Mejstrikova E, Hrusak O, Borowitz MJ, Whitlock JA, Brethon B, Trippett TM et al (2017) CD19-negative relapse of pediatric B-cell precursor acute lymphoblastic leukemia following blinatumomab treatment. jBlood Cancer J 7(12):659
81. Nagel I, Bartels M, Duell J, Oberg HH, Ussat S, Bruckmueller H et al (2017) Hematopoietic stem cell involvement in BCR-ABL1-positive ALL as a potential mechanism of resistance to blinatumomab therapy. Blood 130(18):2027–2031
82. Topp MS, Gokbuget N, Zugmaier G, Klappers P, Stelljes M, Neumann S et al (2014) Phase II trial of the anti-CD19 bispecific T cell-engager blinatumomab shows hematologic and molecular remissions in patients with relapsed or refractory B-precursor acute lymphoblastic leukemia. J Clin Oncol 32(36):4134–4140
83. Topp MS, Gokbuget N, Zugmaier G, Degenhard E, Goebeler ME, Klinger M et al (2012) Long-term follow-up of hematologic relapse-free survival in a phase 2 study of blinatumomab in patients with MRD in B-lineage ALL. Blood 120(26):5185–5187
84. Maury S, Chevret S, Thomas X, Heim D, Leguay T, Huguet F et al (2016) Rituximab in B-lineage adult acute lymphoblastic leukemia. N Engl J Med 375(11):1044–1053
85. Kantarjian HM (2016) Inotuzumab ozogamicin versus standard therapy for acute lymphoblastic leukemia. N Engl J Med 375:740–753

86. Fry TJ, Shah NN, Orentas RJ, Stetler-Stevenson M, Yuan CM, Ramakrishna S et al (2018) CD22-targeted CAR T cells induce remission in B-ALL that is naive or resistant to CD19-targeted CAR immunotherapy. Nat Med 24(1):20–28
87. Kohrt HE, Houot R, Goldstein MJ, Weiskopf K, Alizadeh AA, Brody J et al (2011) CD137 stimulation enhances the antilymphoma activity of anti-CD20 antibodies. Blood 117(8):2423–2432
88. Houot R, Kohrt H, Goldstein MJ, Levy R (2011) Immunomodulating antibodies and drugs for the treatment of hematological malignancies. Cancer Metastasis Rev 30(1):97–109
89. Busch L, Mougiakakos D, Buttner-Herold M, Muller MJ, Volmer DA, Bach C et al (2018) Lenalidomide enhances MOR202-dependent macrophage-mediated effector functions via the vitamin D pathway. Leukemia
90. Grupp SA, Kalos M, Barrett D, Aplenc R, Porter DL, Rheingold SR et al (2013) Chimeric antigen receptor-modified T cells for acute lymphoid leukemia. N Engl J Med 368(16):1509–1518
91. Abramson JS, McGree B, Noyes S, Plummer S, Wong C, Chen YB et al (2017) Anti-CD19 CAR T Cells in CNS Diffuse Large-B-Cell Lymphoma. N Engl J Med 377(8):783–784

9 Minimal Residual Disease in Multiple Myeloma: Impact on Response Assessment, Prognosis and Tumor Heterogeneity

Natalie Berger, Seunghee Kim-Schulze, and Samir Parekh

Abstract

Multiple Myeloma (MM) therapy has evolved rapidly over the past decade. With current multidrug combinations and autologous transplant, rates of overall response exceed 90% and complete response (CR) more than 50% in some studies. Unfortunately, despite higher rates of CR, relapse rates remain high suggesting that persistent disease may not be measured by current techniques. Traditionally, response rates were defined by urine and serum protein electrophoresis, immunofixation and histopathological absence of clonal plasma cells in the bone marrow. Currently, there are several validated sensitive assays to evaluate for MRD (minimal residual disease); multiparameter flow cytometry (MFC) including nextgeneration flow cytometry (NGF), next-generation sequencing (NGS), and allele specific oligonucleotide quantitative polymerase chain reaction (ASO-qPCR). These methods have provided a means to quantitatively assess residual disease and accurately prognosticate PFS and OS in myeloma. In this chapter, we will discuss the current techniques for MRD detection as well as describe techniques that are emerging for improved characterization of drug resistant residual populations that could be adapted for MRD monitoring in the future. While improved therapies are able to eradicate the dominant clone, resistant sub-clones persist and remain undetectable even by MRD techniques. Characterization of these clones will help design therapies against drug-resistant clones and move us closer to a cure in MM.

Keyword

Multiparametric flow cytometry · Multiple myeloma · MRD · Mass cytometry · Next generation sequencing · CD138 · ASO-qPCR · Bone marrow · Microenvironment · ECM

Author contributed equally with all other contributors.
Natalie Berger and Samir Parekh

N. Berger
Hematology/Oncology Fellowship Program, Icahn School of Medicine at Mount Sinai, New York, NY, USA

S. Kim-Schulze
Department of Oncological Science, Icahn School of Medicine at Mount Sinai, New York, NY, USA

S. Parekh (✉)
Division of Hematology and Medical Oncology, Icahn School of Medicine at Mount Sinai, New York, NY, USA
e-mail: samir.parekh@mssm.edu

. A. Aguirre-Ghiso (ed.), *Biological Mechanisms of Minimal Residual Disease and Systemic* *ancer*, Advances in Experimental Medicine and Biology 1100,
s://doi.org/10.1007/978-3-319-97746-1_9

9.1 Introduction

Multiple Myeloma (MM) therapy has evolved rapidly over the past decade. With current multi-drug combinations and autologous transplant, rates of overall response exceed 90% and complete response (CR) more than 50% in some studies [1]. Unfortunately, despite higher rates of CR, relapse rates remain high suggesting that persistent disease may not be measured by current techniques. Traditionally, response rates were defined by urine and serum protein electrophoresis, immunofixation and histopathological absence of clonal plasma cells in the bone marrow. Genomic heterogeneity in myeloma cells manifests in multiple clones of plasma cells. In vitro and animal models demonstrate that MM cells can arise from both immature plasmablasts and mature plasma cells both of which alternate between quiescent and proliferative subtypes [2]. These quiescent cells express a phenotype which is CD138- or low, CXCR4 high, and CD44 positive and also express adhesion molecules and BTK. Matsui et al. used human MM cell lines and NOD/SCID mice to study engraftment of MM cells and better understand the clonal ability of these cells [3]. They demonstrated that human MM cell lines had a small subpopulation of CD138- cells which were more clonogenic than CD138+ cells in vitro and in NOD/SCID mice. Cells which were CD138+ were unable to engraft in mice. This data suggested that MM stem cells are CD138- and have the ability to differentiate into mature CD138+ MM cells. The bone marrow niche plays a key role in supporting quiescent myeloma cells and protects these cells from apoptotic stimuli [4]. This was demonstrated using NOD/SCID mouse models where a fluorescent marker PKH was labeled to MM cells and allowed to cycle in vivo. Quiescence was identified by tracking the intensity of PKH staining, which decreases with each cell division, thereby labeling the quiescent cells within the dividing population of cells. These cells were then tracked in different niches, and quiescent PKH+ MM cells favored forming colonies within the osteoblastic niches of the bone marrow, were highly tumorigenic, and displayed resistance to several drugs [4]. There is greater understanding of interactions between myeloma cells and microenvironment contributing to drug resistance. To identify clonal neoplastic plasma cell populations associated with relapse, more sensitive response techniques have been developed to assess drug resistant populations responsible for minimal residual disease (MRD).

Currently, there are several validated sensitive assays to evaluate for MRD; multiparameter flow cytometry (MFC) including next-generation flow cytometry (NGF), next-generation sequencing (NGS), and allele specific oligonucleotide quantitative polymerase chain reaction (ASO-qPCR) (Table 9.1). These methods have provided a means to quantitatively assess residual disease and accurately prognosticate PFS and OS in myeloma. In this chapter, we will discuss the current techniques for MRD detection as well as describe techniques that are emerging for improved characterization of drug resistant residual populations that could be adapted for MRD monitoring in the future. While improved therapies are able to eradicate the dominant clone, resistant sub-clones persist and remain undetectable even by MRD techniques. Characterization of these clones will help design therapies against drug-resistant clones and move us closer to a cure in MM.

9.2 Current Techniques for MRD Detection

9.2.1 Multiparametric Flow Cytometry (MFC)

Flow cytometry is an important method for diagnosis and monitoring of disease in many hematologic malignancies. Flow cytometry can detect abnormal cell surface and intracellular markers to differentiate between normal and clonal plasma cells. This has led to its application in MM and multiple trials as discussed below have demonstrated a consistent negative prognostic impact of MRD detection by MFC following therapy. The sensitivity of current marker combinations and 8–10 color MFC is 10^{-6}, which

Table 9.1 Comparison of techniques to measure MRD from the bone marrow

Method	Advantages	Disadvantages
Multiparameter flow cytometry	Fast turnaround time (hours)	Lower sensitivity than newer flow methods
	Lower cost	Fresh sample required (ideally << 24 hours)
	Widely available	Not applicable to extramedullary disease
	Applicable to most patients (>95%)	Lacks standardization between individuals and labs
	Quantitative	
	Can detect hemodilution	
	No baseline sample required	
	Sensitive (10^{-5} to 10^{-6})	
Next-generation flow cytometry	High sensitivity (10^{-6})	2-tube 8 color flow is more expensive than MFC
	Faster turnaround time than molecular methods (hours)	Fresh sample required
	Widely available	Not available in all labs
	Applicable to most patients	Not applicable to extramedullary disease
	Quantitative	
	No baseline sample required	
	Standardized between individuals and labs	
Next-generation sequencing	High sensitivity (10^{-6})	Longer turnaround time
	Higher applicability than ASO-PCR (~90%)	High cost
	Does not require fresh sample	Limited availability
	Easy to standardize	Requires baseline sample
		Cannot detect hemodilution
		Indirect quantitation
		Not applicable to extramedullary disease
Allele specific oligonucleotide quantitative polymerase chain reaction	High sensitivity (10^{-5} to 10^{-6})	Longer turnaround time (weeks)
	Reproducible between labs	Labor intensive
	Standardized	Requires baseline sample
	Does not require fresh sample	High cost
		Cannot detect hemodilution
		Not applicable to extramedullary disease
		Not as widely available
		Less applicable (~60–90%)
		Indirect quantitation

means fewer than 1 abnormal plasma cells in one million normal cells is detected (Fig. 9.1). Typically 8 color flow cytometry includes the cell surface markers for CD19, CD45, CD38, CD138 and CD56 as a primary markers followed by any of CD20, CD117, CD200, CD81 and CD27 as secondary markers. Ten color MFC includes the light chain kappa and lambda intracellular detection for the clonality. In addition, the ease of bone marrow aspiration, widespread availability of flow cytometers and rapid turn-around of this test have contributed to the adoption of this technique world-wide for MRD detection in MM.

Myeloma cells have heterogeneous immunophenotypes that allow the distinction of normal from neoplastic plasma cells within the individual patient. When performing flow cytometry analysis, cells are gated by common marker expression, side scatter, and forward scatter to quantify the cells of interest. Normal as well as abnormal plasma cells are primarily gated by CD38 and CD138. CD38 is a type II transmembrane glycoprotein that synthesizes and hydrolyzes cyclic adenosine 5′-diphosphate-ribose, an intracellular calcium ion mobilizing messenger. CD38 is expressed at low levels on the surface of

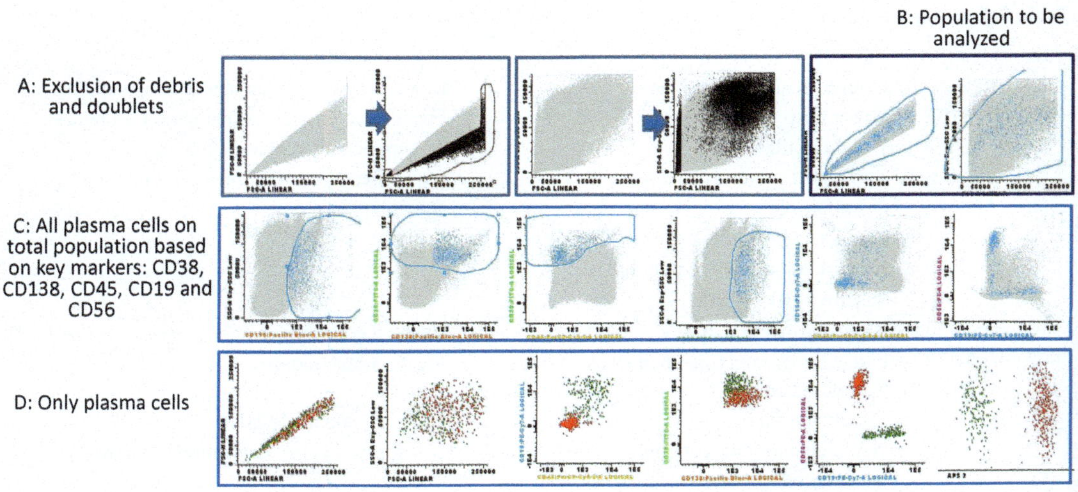

Fig. 9.1 Example to NGF and MRD (Include gating strategy)
Minimal Residual Disease flow cytometry gating strategy. (**a**) Exclusion of non- and cellular debris and doublets by using FSC and SSC plots. Before (grey color total acquisition events) and after (black removed events). (**b**) gated populations for the downstream plasma cell analysis. (**c**) All plasma cells were gated at the background of the total cellular populations. Key makers are CD38, CD138, CD45, CD19 and CD56. (**d**) Only the plasma cells with malignant PC (mPC, red) and normal PC (nPC, green). The last plot shows the distinct separation of the m- and nPC by Automatic Population Separation (APS) function in Infincyte software 2.0

all immune cells and specifically expressed on plasma cells [5]. Cell surface marker CD138 (also known as Syndecan-1) is a proteoglycan also highly expressed in plasma cells. Expression of CD138 decreases during processing of marrows for flow cytometry. Normal plasma cells (nPC) express CD138 and slightly higher levels of CD38 than malignant PC (mPC). nPC expresses CD19, CD45 but do not express CD56 [6]. Malignant clones usually lack expression of CD19 (~95%) and CD45 (~50–70%) and strongly express CD56 (~75%). Analyzing additional antigens which are aberrantly expressed on mPC include CD20 (~15–20% of cases), CD117 (~33% of cases), CD200 (~75% of cases), CD81 (~55%), and CD27 (~40–68%) thus enabling differentiation of normal plasma cells from malignant cells in almost 100% of cases [7, 8]. However, caution should be used as there is variability in antigen expression across each case. In addition, immunophenotypic expression of malignant plasma cells changes after each line of therapy. Notably, newer drugs targeting CD-38 (daratumumab) and CD-138 can result in assay interference by masking the epitopes bound by antibodies used in detection. This is important when evaluating for MRD in patients treated with novel antibody-based regimens.

The direct correlation between MRD with PFS and OS has been demonstrated in several trials (Table 9.2). MRD is assessed at day 100 post autologous stem cell transplant (ASCT) and MRD negativity is associated with improved PFS and OS in MM.

The Spanish myeloma group implemented 4 color flow cytometry using cell surface markers CD38, CD56, CD19, and CD45 to assess MRD in the PETHEMA/GEM2000 study of newly diagnosed multiple myeloma who received standardized induction plus ASCT [9]. MRD status was assessed in 295 patients at day 100 after ASCT. Patients who were MRD negative by MFC limit of 10^{-4} compared to MRD positive had improved median PFS (71 months vs 37 months, $p < 0.001$) and median OS (not reached vs 89 months, $p = 0.002$). The prognostic importance of MRD negativity was also demonstrated in 147 patients who achieved immunofixation negative CR after ASCT. This study demonstrated that MRD status was the most important independent

Table 9.2 Select studies using MRD detection in myeloma

Study	Treatment	N	Method	% of MRD negative patients	Outcomes
Paiva et al. [8]	Newly diagnosed MM from GEM 2000. Six alternating cycles of VBMCP and VBAD then ASCT.	295	4 color MFC	42%	mPFS 71 months vs 37 months (p < 0.001)
					mOS NR vs 89 months (p = 0.002) longer in patients MRD negative day 100 after ASCT
Paiva et al. [45]	Transplant ineligible patients from GEM05 > 65y trial. Six cycles of VMP or VTP.	102	4 color MFC	30%	MRD translated into superior PFS NR vs 35 months (p = 0.02) and TTP compared to conventional CR or sCR but no significant difference in OS.
Korthals et al. [46]	2–4 cycles of idarubicin and dexamethasone then ASCT.	53	ASO-PCR	49%	mEFS in low-MRD group longer than the high-MRD group, 35 mo vs 20 mo (p = 0.001).
					OS longer for low MRD group, 70 mo vs 45 mo (p = 0.04).
Paiva et al. [47]	Patients from GEM2000 and GEM2005MENOS65 who achieve CR at day 100 post ASCT.	241	4 color MFC	64%	High risk cytogenetics and MRD positivity at day 100 post ASCT were the only independent factors that predicted unsustained CR (3 years: TTP 0%, OS 32%; p <.001)
Rawstron et al. [10]	MRC IX trial of newly diagnosed MM: Intensive pathway CTD vs CVAD then ASCT	397	6 color MFC	62%	mPFS MRD negative patients 28.6 mo vs 15.5 mo in MRD positive patients (p < 0.001). mOS 80.6 mo vs 59 mo respectively (p = 0.2)
	MRC IX trial newly diagnosed MM: Non-intensive pathway MP vs CTD	245	6 color MFC	15%	mPFS in MRD positive patients after induction non-significantly inferior to MRD negative patients, 7.4 mo vs 10.5 mo (p = 0.1)
Martinez-Lopez et al. [38]	Induction: VBMC or VBAD then ASCT or 6 cycles of VMP or VTP	133	NGS		mTTP in MRD negative patients 80 mo vs 31 mo (p < 0.0001) in MRD positive patients.
					mOS NR vs 81 months respectively (p = 0.02)
Puig et al. [21]	Patients from GEM2000 and GEM05MENOS65 treated with induction: VBMCP or VBAD then ASCT or 6 cycles of VMP or VTP	170	ASO-PCR	47%	MRD negative compared to MRD positive patients had improved PFS in intensively treated group (54 mo vs 27 mo; p = 0.001) and in the non-intensively treated group (median NR vs 31 mo; p = 0.029); OS not significantly different in either group.

(continued)

Table 9.2 (continued)

Study	Treatment	N	Method	% of MRD negative patients	Outcomes
Roussel et al. [48]	3 induction cycles of RVD then ASCT followed by 2 RVD consolidation cycles and 1 year of Len maintenance.	31	7 color flow cytometry	68%	MRD negative patients had an estimated 3 year relapse free survival of 100%.
Ferrero et al. [49]	Updates results from GIMEMA-VEL-03-096 trial. Patients treated with VTD then ASCT.	39	ASO-PCR		At 8 years OS in MRD negative patients 72% vs 48% in MRD positive patients ($p = 0.041$).
Korde et al. [50]	Newly diagnosed MM and high risk SMM 8 cycles of KRD followed by Len maintenance if stable disease or better.	45 (MM)	MFC and NGS	N = 28 achieved CR or better, 100% MRD negative by MFC and 67% by NGS	PFS 100% vs 79% at 12 months in MRD negative vs positive patients ($p < 0.001$)
		12 (SMM)			
Ludwig et al. [51]	Newly diagnosed MM VTD vs VTD-C followed by ASCT in transplant eligible patients.	98 (total)	4 color MFC	80% of BM CR	mPFS NR vs 38.6 months in the MRD negative vs positive patients ($P = 0.085$).
		42 (evaluated for MRD)			mOS NR in either group ($p = 0.032$).
Avet-Loiseau et al. [44]	Patients with relapsed/refractory MM on POLLUX (DRd vs RD) and CASTOR (DVd vs Vd).	POLLUX: 3% in DRd and 87% in Rd. did not receive MRD assessment	NGS	Varied based on sensitivity threshold	The addition of daratumumab resulted in higher MRD negative rates.
		CASTOR: 76% in DVd and 87% in Vd did not receive MRD assessment.			At sensitivities of 10^{-6}:
					POLLUX 9.8% vs 2.1% in DRd vs Rd. ($p = 0.00006$).
					CASTOR 3.6% vs 0.8% in DVd vs Vd ($p = 0.028$).
Oliva et al. [52]	Patients from RV-MM-EMN-441 study who received induction + VMP, CRD, or ASCT followed by len maintenance	50	MFC	38% after consolidation	After consolidation 5-year PFS was 72% for MRD negative vs 30% for MRD positive patients ($p = 0.005$)
				Additional 14% during maintenance	During maintenance mPFS in MRD was NR for MRD negative vs 35 months for MRD positive patients ($p = 0.0004$)
					5-year PFS was 94% for MRD negative vs 33% for MRD positive patients at 12 months of maintenance ($p < 0.001$)

Paiva et al. [53]	Transplant ineligible patients from PETHEMA/GEM2010MAS65 alternating VMP plus Rd	162	NGF	34%	MRD monitoring is one of the most relevant prognostic factors in elderly MM patients, irrespectively of age or cytogenetic risk
					MRD negative patients mTTP NR vs 15 months in MRD positive patients.
					3 year OS 70% in MRD negative patients vs 55% in MRD positive patients.
de Tute et al. [11]	Patients from MRC myeloma IX trial treated with CTD or CVAD followed by ASCT +/− len maintenance	397	6 color MFC	End of induction: CTD 25%, CVAD 13%	Post ASCT mPFS for MRD negative vs MRD positive patients is 28.6 vs. 15.9 months, p < 0.001; median OS 80.6 vs. 59.0 months, p = 0.018
				Post ASCT 71% vs. 54% respectively	
Attal et al. [54]	RVD +/− ASCT followed by RVD plus len maintenance.	543	7 color MFC (lambda/CD56 + CD28/CD138/CD19/kappa/CD45/CD38)	65% in RVD alone group	PFS was longer in MRD negative vs MRD positive patients (adjusted hazard ratio, 0.30; 95% confidence interval, 0.23 to 0.37; p < 0.001).
				79% in the transplant group	OS was longer in MRD negative vs MRD positive patients (adjusted hazard ratio, 0.34; 95% confidence interval, 0.22 to 0.51; p < 0.001).

Abbreviations: *MRD* minimal residual disease, *MM* multiple myeloma, *SMM* smoldering multiple myeloma, *VBMCP* vincristine, carmustine, melphalan, cyclophosphamide, and prednisone, *VBAD* vincristine, carmustine, adriamycin, dexamethasone, *MFC* multiparametric flow cytometry, *ASCT* autologous stem cell transplant, *mPFS* median progression free survival, *mOS* median overall survival, *VMP* bortezomib, melphalan, and prednisone, *VTP* bortezomib, thalidomide, and prednisone, *NR* not reached, *TTP* time to progression, *CR* complete response, sCR stringent complete response, *mEFS* median event free survival, *CTD* cyclophosphamide, thalidomide, and dexamethasone, *CVAD* cyclophosphamide, vincristine, Adriamycin, and dexamethasone, *MP* melphalan and prednisone, *CTD* cyclophosphamide, thalidomide, dexamethasone, *VBMC* vincristine, carmustine, melphalan, cyclophosphamide, *RVD* lenalidomide, bortezomib, dexamethasone, *Len* lenalidomide, *VTD* bortezomib, thalidomide, dexamethasone, *KRD* Carfilzomib, lenalidomide, dexamethasone predictor of PFS (HR =3.64, $p = 0.002$) and OS (HR = 2.02, $p = 0.02$).

This PFS and OS benefit was also demonstrated in the Spanish GEM2000 and GEM 2005 trials in a preplanned subgroup analysis. This study demonstrated that the best independent predictors for early relapse after achieving CR were presence of MRD day 100 after ASCT by 4 color flow cytometry or presence of high risk cytogenetics by FISH. Early relapses after achieving CR were associated with a poor survival.

The Medical Research Council (MRC) myeloma IX trial included patients with newly diagnosed MM who received intensive therapy pathway or non-intensive pathway [10]. The intensive pathway was either CTD (cyclophosphamide, thalidomide, and dexamethasone; n = 189) or CVAD (cyclophosphamide, vincristine, doxorubicin, and dexamethasone; n = 208) for 4–6 cycles followed by high dose melphalan 200 mg/m2 and ASCT [10–12]. There was also a non-intensive pathway in which patients received melphalan and prednisolone (MP) vs. Attenuated CTD (CTDa) (N = 856) for 6–9 cycles. Patients were then randomized to thalidomide maintenance (50 or 100 mg daily) or no further treatment. In the intensive cohort absence of MRD by 6 color flow cytometry (using cell surface markers CD138, CD45, CD38, CD19, CD56, CD27) at day 100 was predictive of an improvement in PFS (28.6 vs 15.5 months; $p < 0.001$ in the MRD negative compared to MRD positive patients) and OS (80.6 vs 60 months; $p = 0.02$ respectively). In the non-intensive cohort 245 patients (29%) achieved a CR, and of those, MRD positivity at the end of induction was associated with a non-statistically significant inferior PFS (7.4 vs 10.5 months, $p = 0.1$) and nonsignificant inferior OS. The impact of maintenance thalidomide on PFS was also evaluated in an unplanned subgroup analysis. MRD positive patients that did not receive maintenance had the shortest PFS, whereas the longest PFS was seen in patients who were MRD negative and received maintenance therapy ($p < 0.001$), and 28% of MRD positive patients became MRD negative with maintenance.

MRD testing in these studies was pre-planned but carried out as subgroup analyses which may limit the power as well as introduce selection bias. In addition, the sensitivity and specificity of the flow methods used in these trials are inferior to the currently available 8 color (only cell surface markers) and 10 color (includes the light chain kappa and lambda) flow cytometry. The currently available 2 tube 8 color flow cytometry is standardized across all laboratories making interpretation more uniform.

Multiple studies using 8 color flow have reported the correlation of MRD negativity with an improved PFS and OS. A meta-analysis of MRD status on PFS or OS in 20 or more patients [13], demonstrated the prognostic value of MRD on PFS with a total of 1273 patients; 660 MRD negative and 613 MRD positive. The impact of MRD on OS was evaluated in 12 studies which included 1100 patients of which 599 were MRD negative and 501 were MRD positive. Results for patients who achieved a conventional CR were reported in 5 studies (n = 574) for PFS and 6 studies (n = 616) for OS. MRD negative status compared to MRD positive status was associated with a better PFS (HR 0.41; 95% CI, 0.36–0.46; $p < 0.001$) and OS (HR 0.57; 95% CI, 0.46–0.71; $p < 0.001$). Median PFS was 54 months for MRD negative patients and 26 months for MRD positive patients. Median OS was 98 vs 82 months respectively. Tests for heterogeneity found significant differences in studies for PFS and OS.

MRD negativity after ASCT retains prognostic significance on outcomes in patients with both standard and high risk cytogenetics. The myeloma IX study included patients with unfavorable cytogenetics (defined as gain(1q), del(1p32), t(4;14), t(14;20), t(14;16), and del(17p)). Regardless of cytogenetic risk group, absence of MRD at day 100 post ASCT was associated with an improved PFS (< 0.001) and OS ($p = 0.0183$) [10].However, the median PFS was three times longer in MRD negative standard risk patients compared to MRD negative high risk patients. This suggests that the disease drivers identified by cytogenetic analysis contribute significantly to prognosis not just at diagnosis, but during disease progressing after treatment i.e. at relapse in the MRD negative population.

The impact of post ASCT stringent complete response (sCR) and MRD status (using 6 color MFC with cell surface marker CD19, CD38, CD45, CD138, kappa, lambda) on PFS and OS was retrospectively studied in 185 consecutive patients with high risk cytogenetics [14]. High risk cytogenetics was defined as deletion (17p), t(4;14), t(14;20), t(14;16), or + 1q by FISH. Patients who achieved sCR had improved PFS compared to those who did not (median, 38 vs 21 months, $p = 0.002$). MRD negativity was achieved by 103 patients (56%) on day 100 post ASCT. Patients who were MRD negative had an improved PFS compared to those who were MRD positive (median, 26 vs 17 months, $p < 0.001$) and superior OS (5 year OS 64% vs 41%, $p = 0.023$). In subgroup analysis, sCR and MRD negativity did not translate into improved PFS or OS in patients with deletion (17p) or ≥ 2 HR abnormalities. In patients with t(4;14), post ASCT MRD negativity improved PFS (24 vs 12 months, $p = 0.002$) and OS (NR vs 31 months, $p = 0.039$) compared to those who were MRD positive. Patients with high risk cytogenetics who achieve sCR or MRD negativity after ASCT have improved PFS and OS, except patients with deletion (17p) or ≥2 HR cytogenetic abnormalities. The addition of post-transplant maintenance with PIs and/or IMIDS has resulted in improved PFS and a trend towards improved OS. The most sensitive flow cytometry methods available now include next generation flow (NGF) 2-tube 8 color flow cytometry and single tube 10 color flow cytometry. The EuroFlow NGF method is part of the Black Swan Initiative (BSI) in an effort to standardize MRD testing and improve sensitivity. 2-tube 8-color antibody panel uses two tubes; one tube of surface antigens to identify myeloma cells and a second tube to test cytoplasmic kappa and lambda light chain expression [7]. This method also has a standardizing software tool (Infinicyte2.0) for automatic plasma cell gating (Fig. 9.2). Figure 9.2 shows the representative gating strategy for the nPC and mPC based on the Automated Population Separator Algorithm used in Infinicyte 2.0. This method was validated in a multicenter analysis comparing conventional 8 color flow cytometry to NGF method in 228 BM samples from myeloma patients at diagnosis ($n = 66$) or after therapy ($n = 162$). Of these, 110 samples were from patients with very good partial response (VGPR) or CR/sCR. NGF compared to conventional 8 color flow showed higher sensitivity, with an MRD positive rate of 47% vs 34% ($p = 0.003$). Therefore, 25% of patients who were MRD negative by conventional flow were MRD positive by NGF. This with NGF this translated into longer PFS for MRD negative vs MRD positive CR (75% PFS NR vs 7 months, $p = 0.02$). They were also able to consistently measure $>5 \times 10^6$ events per patient sample which likely explains the improved sensitivity by conventional flow cytometry.

Single 10 color tube antibody combinations have also been developed. A direct comparison between 2-tube 8 color flow and 10 color flow was performed and revealed comparable results. Differences included a higher number of cells measured by 2-tube 8 color flow with less cell loss in the first tube cell surface staining and the added benefit of a second confirmatory tube [15, 16]. Benefits to the single 10 color tube method include lower cost due to reduced labor, reagents, and processing time.

While MFC has many advantages, it does have its limitations and critically depends on bone marrow aspiration quality. False negative results from MFC can be due to a number of factors including hemodilution, sampling error, plasma cell fragility, and limitations of flow cytometry as discussed below. Unlike leukemic cells which can survive ex-vivo, myeloma cells are less resilient and die rapidly when removed from the bone marrow. As a result, false negative results should be interpreted with caution.

Multiple myeloma cells are known to be underrepresented in bone marrow aspirates. This appears to be multifactorial from fibrosis and adhesion, hemodilution, and rapid cell death after removal from the bone marrow [17, 18]. Myeloma cells promote cancer associated fibroblasts (CAFs) resulting in reticulin fibrosis encasing myeloma cells. This interaction enables tumor

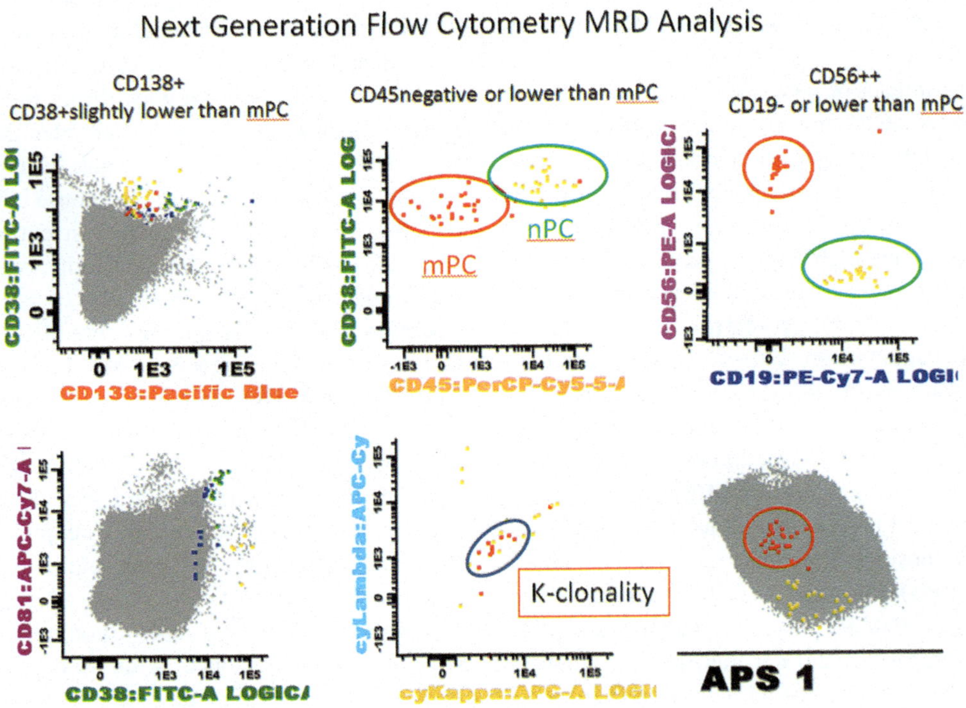

Fig. 9.2 Representative example minimal residual disease assay via next generation flow cytometry method and analysis using infinicyt 2.0 software
Representative example of Minimal Residual Disease assay via Next Generation Flow Cytometry method and analysis using Infinicyt 2.0 software. Normal PC (nPC) and Malignant PC (mPC) populations from a representative bone marrow aspirate were identified using the Euroflow 8 surface markers and intracellular Kappa and Lambda cytoplasmic expression. In general, mPC express slightly low CD38 and CD45 low/negative expression compared to nPC as shown in the top panel middle CD45 (x-axis) and CD38 (y-axis) dot plots. mPC is predominantly CD56 positive while nPC is CD56 negative. This patient mPC express cytoplasmic kappa light chain restriction with expression of CD56 without CD45 or CD19. Automatic Population Separation (APS) is PCA based clustering of the m- and nPC for unbiased clustering of the PCs based on the weighted 10 antigen expressions

progression by supporting cellular proliferation, extracellular matrix (ECM) remodeling, resistance to apoptosis, and promoting angiogenesis. The ECM encasing myeloma cells makes aspiration of malignant plasma cells more challenging and preferentially leads to aspiration of other cells within the bone marrow [17]. In addition, myeloma cells express adhesion molecules such as CD56 and adhere to the stroma. Several studies have showed that myeloma cells are underrepresented on aspirates. In a study by Joshi et al. bone marrow aspirates showed plasma cell percentage far less than seen on bone marrow biopsy (13.1% vs 31.8% respectively) [19]. Traditionally during a bone marrow aspirate the first pull is reserved for morphology slides. As more blood is aspirated from the marrow it will inevitably be contaminated by peripheral blood [20]. Consequently, an aliquot of the first aspirate should be sent for assessment of MRD to minimize the risk of a false negative result. Assessment for MRD is also limited due to rapid cell death once myeloma cells are removed from the bone marrow and specimens should be processed within a few hours.

9.2.2 ASO-qPCR

There are two molecular methods for detecting MRD in MM; allele-specific oligonucleotide quantitative polymerase chain reactor

(ASO-qPCR) and next generation sequencing (NGS). Polymerase chain reaction for monitoring MRD in multiple myeloma can be done using an allele-specific oligonucleotide (ASO) which can identify clonal plasma cell specific immunoglobulin heavy chain gene rearrangements. This method is very specific and sensitive detecting up to 10^{-5} cells. It requires a baseline sample when disease burden is high to produce patient specific probes which are complementary to the junctional region of the rearranged IGH genes. These probes can then identify residual disease in the bone marrow at different time points. Due to high number of somatic hypermutations its applicability is more restricted to 50–90% of cases [21, 22].

ASO-qPCR and MFC have been compared in several studies. In a study by Puig et all [21] these 2 methods were compared in a series of 170 patients who achieved at least a partial response and had bone marrow samples available (at diagnosis and after treatment) from 3 consecutive Spanish PETHEMA/GEM trials. The average percentage of residual myeloma cells detected by ASO-qPCR and four-color MFC was not significantly different (0.31 vs 0.39, respectively). Residual myeloma cells were detected by ASO-qPCR in 53% of cases while in 46% of patients using MFC. This suggests a higher sensitivity in the molecular subgroup however, with improved sensitivity of MFC with ≥8 colors, this benefit may no longer be seen. In addition, there was a significant correlation between the methods ($r = 0.881$; $p < 0.001$) in detecting MRD. However, 58% of patients could not be evaluated by ASO-qPCR due to absence of clonality, unsuccessful sequencing, or suboptimal ASO-qPCR performance.

While ASO-qPCR was formerly considered the best method at detecting MRD due to its higher sensitivity and ability to use banked samples, subsequent developments in MFC have led to increased popularity of the MFC method due to its widespread availability, improving sensitivity, improved standardization, lower cost, and rapid processing time.

9.2.3 Next Generation Sequencing (NGS)

Next-generation sequencing is a highly sensitive molecular technique to measure MRD in the bone marrow detecting up to 10^{-6} cells [23]. Most data on NGS is using the LymphoSIGHT platform (Sequenta Inc). This platform uses sets of multiple primers to amplify and sequence immunoglobulin gene segments with locus specific primers for IGH-VDJH, IGH-DJH, or IGK rearrangements. After amplification immunoglobulin gene DNA is sequenced to determine the frequencies of different clones. Baseline samples with >5% myeloma clones can be studied in the future to measure MRD at different time points during treatment (Fig. 9.3).

NGS has been compared to ASO-qPCR and MFC. ASO-qPCR has the limitation of requiring patient specific probes which is not required by NGS. These two methods were compared in patients with MM, MCL, and ALL to determine if NGS could overcome some limitations of ASO-qPCR and further increase sensitivity, specificity, accuracy, and reproducibility [24]. Clonotypes detected by both methods were identical or >96% homologous in 96% of cases. Correlation of MRD results was good ($R = 0.791$, $p < 0.001$) and concordance in 79.6% of cases. Both tests reached sensitivities of 10^{-6}.

9.2.4 Imaging Techniques for MRD (PET-CT and MRI)

The MRD assays mentioned above rely on random sampling of tumor cells from bone marrow aspiration from the iliac crest. Myeloma patients have multifocal disease with spatial heterogeneity in clones [25]. To reduce the false negative rate due to sampling error, the IMWG has revised the criteria for complete remission to include MRD and PET negative CR [12]. PET/CT can detect extramedullary disease (Fig. 9.4) and can also detect focal areas of bone marrow infiltration by MM cells which may be missed by blindly performing a bone marrow biopsy. This technique

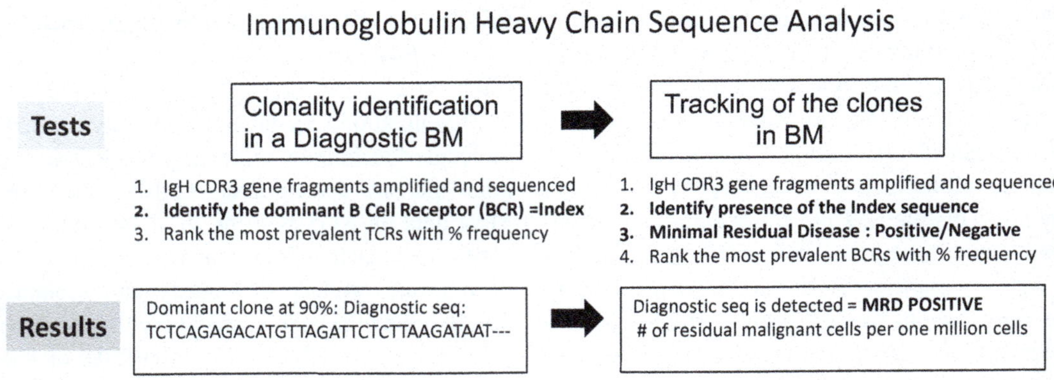

Fig. 9.3 Schematic representation of steps required for the immunoglobulin heavy chain sequence analysis Schematic representation of Steps required for the immunoglobulin Heavy Chain Sequence Analysis. Tests: Diagnostic initial bone marrow aspirate DNA is subjected to the immunoglobulin heavy chain sequence analysis to identify the dominant B cell receptor clone (=Index). Nucleotide sequences of the dominant clone with the percent frequency (i.e. 90%) and the rest of the prevalent TCRs are reported. Index clones in the follow up bone marrow aspirates are tracked by DNA sequencing to determine the detection of the minimal residual as positive or negative and the number of MRD cells

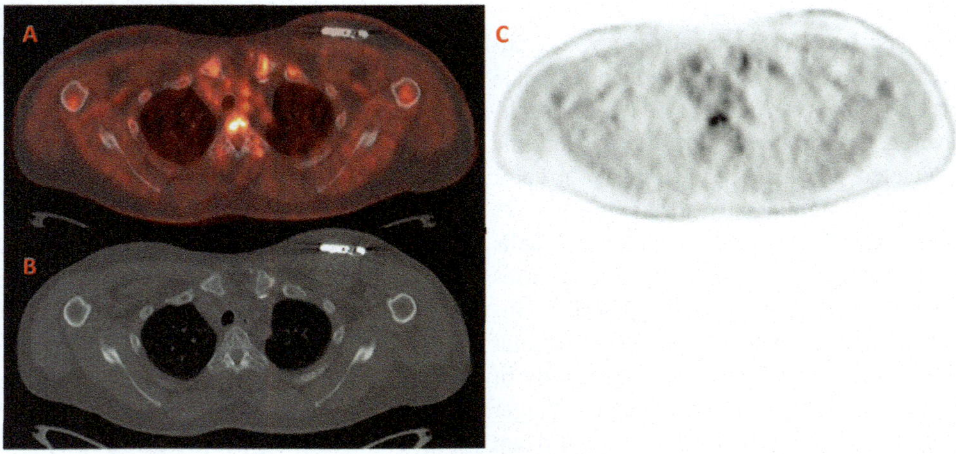

Fig. 9.4 Positive PET/CT in patient who is MRD negative by 2-tube 8 color MFC (**a**) Fused FDG PET/CT axial image demonstrating new increased uptake with SUV max of 4.5. (**b**) Low dose axial CT image showing a lytic lesion in T2 vertebral body. (**c**) PET only image with mild uptake in the region of T2 vertebral body

can be coupled with flow and molecular based techniques to identify MRD at intramedullary and extramedullary sites.

Several trials have shown the prognostic and predictive benefit of PET-CT. The IFM2009 trial included 134 patients who had a PET/CT scan and MRI of the spine and pelvis to compare these modalities for detecting bone lesions [26]. Patients were treated with induction RVD (lenalidomide, bortezomib, dexamethasone), with or without ASCT, followed by lenalidomide maintenance. Imaging was done at diagnosis, after 3 cycles of RVD, and before maintenance treatment. At diagnosis MRI and PET-CT were positive in majority of patients (95 vs 91% respectively). MRI and PET-CT were equally efficacious at detecting bone involvement in symptomatic patients at diagnosis. After 3 cycles of induction, PET-CT normalized in 32% of patients (43/134) and this group of patients had an improved PFS compared to those with a positive PET-CT however, there was no difference in OS. Before maintenance, PET-CT normalized in 62% (83/134) of patients who were positive at baseline, which correlated with an improved PFS and OS. Independent prognostic markers on PFS and OS included extramedullary disease at baseline. PET-CT normalization prior to maintenance was an independent prognostic marker for PFS. After 3 cycles of induction, MRI normalized in 3% of patients and before maintenance it normalized in 11% of patients. This did not result in an improved PFS or OS. This study demonstrated that PET-CT is a valuable tool for evaluation of treatment response and prognosis. The benefit to MRI on assessment of MRD remains unclear.

In a study by Zamangi et al., 192 patients with newly diagnosed MM had a PET/CT at diagnosis, after treatment with thalidomide and dexamethasone (TD), and after double autotransplantation. Patients who had PET/CT at diagnosis with 3 or more focal lesions, SUV (standardized uptake values) >4.2, and extramedullary disease (EMD) had worse 4 year PFS. Those with SUV >4.2 and EMD had worse 4 year OS. Patients with SUV >4.2 after TD had a shorter PFS. PET/CT performed 3 months after ASCT was negative in 65% of patients. This correlated with improved 4 year estimates in PFS (47% vs 32%, $p = 0.02$) and OS (79% vs 66%, $p = 0.02$) compared to those who were PET/CT positive [27].

Patients treated with Total Therapy 3 (TT3) ($n = 239$) who had more than 3 FDG-avid focal lesions prior to ASCT had inferior OS and event free survival (EFS) [28]. Those with complete suppression of FDG-avid lesions had improved outcomes. These outcomes were only opposed in patients with high risk gene expression profiles. This was again demonstrated by Usmani et al. in 302 patients of whom 277 had gene expression profiles. Based on multivariate analysis, more than 3 FDG-avid focal lesions on cycle 1 day 7 of induction chemotherapy was predictive of an inferior PFS and OS [29]. Patients part of TT4–6 ($n = 596$) had a baseline PET/CT and those with >3 focal lesions at presentation had worse PFS and OS. Those patients who had complete improvement in the SUV at day 7 had the same prognosis as those without any lesions on baseline imaging. Even as these patients were serially followed, suppression of SUV signal in focal lesions later in treatment correlated with improved prognosis [30].

A newer imaging technique PET-MRI is available at limited centers and its role in myeloma is under investigation. A comparison of PET/CT and PET/MRI was done in a study by Sachpekidis et al. which included 30 patients with MM who underwent PET/CT followed by PET/MRI [31]. Data was assessed qualitatively and quantitatively based on lesions detected and SUV respectively. Both techniques demonstrated equivalent performances in detecting myeloma skeletal lesions (94.2% of lesions depicted in PET/CT were also seen on PET/MRI). There was a statistically significant difference in lesional SUVs (higher in PET/CT compared to PET/MRI) however, there was a significant correlation in SUV average ($r = 0.744$, $p < 0.0001$) and SUV max ($r = 0.855$, $p < 0.0001$) between both techniques. An additional benefit to PET/MRI includes limiting radiation exposure. Ongoing studies are underway to evaluate the sensitivity of imaging studies compared to flow cytometry.

9.3 Emerging Techniques for Detection of Clonal Populations in Myeloma

Multiple myeloma cells are heterogeneous and drug resistant populations arise during different phases of therapy. The mechanisms behind chemoresistance include A) molecular changes due to treatment and B) tumor heterogeneity where the dominant clone is eradicated during therapy but a resistant and quiescent subclone persists but inevitably proliferates resulting in relapsed disease [32, 33].

Studies have shown that clonal selection due to treatment of myeloma may already be present at the MRD stage and that MRD plasma cells have distinct phenotypes with different cytogenetic and gene expression profiles [33]. A small subset of patients from the GEM2010MAS65 clinical trial who were elderly and transplant ineligible with newly diagnosed myeloma were prospectively studied during front line therapy. The antigenic profile of MRD clonal plasma cells vs diagnostic clonal plasma cells were compared. The study showed that MRD subclones overexpressed integrins, chemokine receptors, and adhesion molecules such as CXCR4 and CD44. Genetic profiling of a subset of these patients from the MRD vs. diagnostic sample showed that there is frequent clonal tiding of genetically distinctive subclones. Another study showed that the genomes of patients with high risk cytogenetics have significantly more changes over time compared to those with standard risk in MM [34]. The results of this study indicate that tumors may be genetically stable, linearly evolving, or have clonal heterogeneity where the dominant clone is variable. Characterizing these clones may help us design treatment protocols which can target resistant clones during frontline therapy.

9.3.1 Time of Flight Mass Cytometry (Cytof)

The current flow cytometry systems allow 10 colors to be analyzed efficiently, however there is a challenge with larger number of marker combinations that can be used simultaneously due to spectral overlap between channels and need for compensation. With the advent of newer therapeutic targets such as the B-Cell Maturation Antigen (BCMA) and others, there is a need to incorporate additional markers to characterize drug resistant residual populations for therapeutic planning.

Next-generation mass cytometry by time of flight (cyTOF), combines flow cytometry with time-of-flight mass spectrometry (MS) instead of fluorescence. This technology uses antibodies labeled with stable heavy metal isotopes which allows for simultaneous assessment of large panels of markers.

Studies examining small numbers of patients suggest a potential of improved sensitivity using cyTOF. In a study of 5 patients with MM, 3 patients were MRD positive by cyTOF when MFC was negative [35]. Using this method rare myeloma cells could be detected and also characterized phenotypically distinct clonal populations of plasma cells. The capacity for studying larger panels of markers can allow characterization of both myeloma cells and their microenvironment, helping assess responses to novel therapeutics and increase our understanding of mechanisms of resistance.

9.3.2 Liquid Tumor Biopsy and Measurement of Circulating Tumor Cells (CTCs)

Serial measurements of MRD are limited as repeat bone marrow biopsies are invasive and painful. Developing a method to detect residual myeloma cells without performing a bone marrow biopsy is an important tool currently being studied by measuring myeloma cells in the peripheral blood. Detection of myeloma cells in the blood is challenging due to their typical restriction to the bone marrow. However, myeloma cells have been detected in the blood using highly sensitive techniques including MFC and ASO-qPCR.

The limitations of incorporating CTC measurement into clinical trials has been the low sensitivity of CTC measurement using current methodologies. Mithraprabhu et al. developed a method to isolate single CTCs even when detection is low by flow cytometry [36]. They combined $CD138^{+}CD45^{-}$ cell enrichment plus serial dilution and single-cell manipulation using fluorescence microscopy of these live cells. This approach resulted in a sensitivity of 100% in 24 patient samples yielding at least 12 CTCs per patient. They were able to detect single myeloma cells in PB and BM with similar sensitivity to MRD diagnosis by fluorescence-activating cell sorting (FACS) ($\leq 1{:}10^5$). In addition, they were able to perform molecular analysis on these CTCs. The ability to monitor the genetic evolution of myeloma cells could enable us to understand mechanisms behind drug resistance, identify and follow acquired mutations during therapy, and even identify patients before they progress from MGUS to MM.

The prognostic value of CTC quantification compared to BM MRD measurement has been studied in several trials. Huhn et al. compared MRD from the BM compared to CTC quantification from PB at diagnosis, after induction, ASCT, consolidation, and maintenance. Sensitivity reached 10^{-6} for MRD measurements from the BM and CTCs. Detection of CTCs was sensitive in patients who were MRD positive with 19/20 PB samples with detectable MM cells also having a positive paired BM. However, in 19/34 pairs, PB was negative for CTCs but paired BM specimens were positive for myeloma cells. Patients who had CTCs in PB after ASCT had worse OS compared to those without CTCs ($p = 0.023$). This further emphasizes the need for highly sensitive detection methods for CTC quantification. The routine feasibility of this method heavily depends on the single cell sequencing technology and has a great potential to be paired with bone marrow MRD assessment.

9.3.3 Mass Spectrometry (miRAMM)

Monoclonal immunoglobulin Rapid Accurate Mass Measurement (miRAMM) is a new technique and can achieve sensitivities up to 10^{-7}. This technique uses mass spectrometry to improve the analytical range of M-protein detection from peripheral blood in patients with MM compared to electrophoresis. Benefits of this method include use of peripheral blood as compared to a more invasive bone marrow biopsy, cost effectiveness, and rapid processing time. The study included 30 MM patients who achieved sCR post ASCT [37]. Samples were assessed at the time of diagnosis, day 100 post ASCT, 6 months, and 12 months. Day 100 post ASCT M-protein was detected by miRAMM in 81% of patients. Patients with PR pre-ASCT and IgG isotypes had the highest rate of M-protein detection. Detection of an M-protein at day 100 post-ASCT, 6 months, or 12 months did not correlate with PFS. Of the sCR patients, those with decreasing or negative M-protein intensities had a longer PFS than those who had stable or increasing M-protein intensities (mPFS 51.6 vs 17.9 months, $p < 0.0017$ respectively). While this is not a substitute for MRD testing by MFC, NGS, or ASO-qPCR, this is an important method to serially follow patients and aid in determining ideal timing for MRD testing.

9.4 Conclusions and Future Directions

Advances in targeted therapies, understanding of the tumor microenvironment and the immunotherapeutics has improved outcomes for myeloma patients. With the identification of MRD as an important component of achieving a CR, classical methods such as immunofixation and the bone marrow plasmacytosis have been demonstrated to be inadequate markers of disease. Patients are achieving MRD negativity at increasing rates and studies have confirmed its correlation with improved PFS and OS. Flow cytometry and molecular methods continue to improve in sensitivity, but each assay for MRD testing has limitations in its current form. This suggests that parallel testing approaches incorporating bone marrow based assays like MFC, PET imaging and serology will be necessary for comprehensive disease assessment, especially in cases where results are discordant [38, 39].

The timing of MRD monitoring should be also carefully assessed. The typical timing for MRD assessment has been at the end of 6–8 cycles of therapy, after induction, at day +100 after autologous stem cell transplant, and during maintenance therapy in various trials (Table 9.2). A strategy examining serial MRD measurements to detect the trajectory of disease before clinical relapse may be a useful approach. Standardization of the timing of MRD assessment, particularly as an early readout of efficacy among clinical trials will be important for future trials to establish.

MRD detection in the bone marrow is limited such that when patients relapse outside of the bone marrow with a genetically disparate clone, only PB CTC single cell analysis would be useful. It is highly anticipated to see whether MRD monitoring in the peripheral blood will be able to capture clonal diversity via single cell sequencing [40].

Single cell sequencing and functional studies are important to better understand the the biology of the resistant clone beyond the technological resolution by bulk sequencing techniques. Single cell sequencing can detect individual cells responsible for sub-clonal evolution and proliferation. The sensitivity of single-cell techniques continues to improve, detecting up to 10,000 cells from an individual tumor. Identifying these sub-clones and understanding tumor heterogeneity within an individual patient will allow for more targeted treatment plans [41].

The next generation of trials will address the use of MRD monitoring to guide clinical decision making. Critical questions around therapy escalation, duration of maintenance therapy or its early discontinuation are under evaluation. A particular example of this is NCT02389517. This study has a primary objective to determine the rate of MRD negative disease by MFC after therapy at 12 months with Lenalidomide with or without Ixazomib and Dexamethasone in patients with residual MM after donor stem cell transplant.

Another approach is studying intensified treatment in patients with PET positivity after first line treatment. The study combines PET/CT and MFC based MRD evaluation, i.e. patients will have a PET/CT after standard first line treatment and those who are PET-positive will have a BM biopsy for MRD evaluation by flow cytometry. MRD negative patients will be excluded and those who are MRD positive then receive 4 cycles of Carfilzomib, Revlimid, Dexamethasone (KRd) followed by a new PET/CT and MRD evaluation by flow cytometry (NCT03314636).

MRD is also being evaluated as an early endpoint for assessing efficacy of novel therapy towards rapid approval e.g. for CAR-T trials in MM. In the phase III CASTOR [42] and POLLUX [43] trials, Daratumumab was combined with bortezomib or lenalidomide respectively plus dexamethadone vs boreztomib or lenalidomide plus dexamethasone in patients with relapse/refractory MM. Patients in the daratumumab arm had improved PFS including those with high risk cytogenetics. MRD was assessed in both trials using NGS and 3 different sensitivity thresholds of 10^{-4}, 10^{-5}, and 10^{-6} [44]. Among patients who achieved CR or better, daratumumab containing regimens reached rates of MRD negativity at least 3 times higher compared with the control groups.

The rapid advancement of technology of quantification and characterization of the rare malignant plasma cells with the sensitivity currently up to 10^{-6} strongly suggest that MRD assay is well suited for monitoring of Multiple myeloma patients disease course. Both molecular (ASO-qPCR and NGS) and cellular (Flow Cytometry) based assays showed efficient applications for the detection of the MRD. These MRD assays are also likely to prove to be advantageous for clinical decision makings. It remains to be seen which of these methods will be most effective. However, questions remain regarding quality harmonization of the bone marrow, biopsy specimen, and standardization of MFC assay methods including the data analysis.

The efficient application of MRD assays and improved characterization of tumor cells using technological advances in non-invasive tumor burden monitoring by liquid biopsy or mass spectrometry will advance the science of MRD towards therapeutic targeting of quiescent drug resistant clones to cure MM.

References

1. Barlogie B, Mitchell A, van Rhee F, Epstein J, Morgan GJ, Crowley J (2014) Curing myeloma at last: defining criteria and providing the evidence. Blood 124(20):3043–3051
2. Yaccoby S (2018) Two states of myeloma stem cells. Clin Lymphoma Myeloma Leuk 18(1):38–43
3. Matsui W, Huff CA, Wang Q, Malehorn MT, Barber J, Tanhehco Y et al (2004) Characterization of clonogenic multiple myeloma cells. Blood 103(6):2332–2336
4. Chen Z, Orlowski RZ, Wang M, Kwak L, McCarty N (2014) Osteoblastic niche supports the growth of quiescent multiple myeloma cells. Blood 123(14):2204–2208

5. Rawstron AC, Orfao A, Beksac M, Bezdickova L, Brooimans RA, Bumbea H et al (2008) Report of the European myeloma network on multiparametric flow cytometry in multiple myeloma and related disorders. Haematologica 93(3):431–438
6. Paiva B, Vidriales MB, Rosinol L, Martinez-Lopez J, Mateos MV, Ocio EM et al (2013) A multiparameter flow cytometry immunophenotypic algorithm for the identification of newly diagnosed symptomatic myeloma with an MGUS-like signature and long-term disease control. Leukemia 27(10):2056–2061
7. Flores-Montero J, Sanoja-Flores L, Paiva B, Puig N, Garcia-Sanchez O, Bottcher S et al (2017) Next generation flow for highly sensitive and standardized detection of minimal residual disease in multiple myeloma. Leukemia 31(10):2094–2103
8. Muccio VE, Saraci E, Gilestro M, Gattei V, Zucchetto A, Astolfi M et al (2016) Multiple myeloma: new surface antigens for the characterization of plasma cells in the era of novel agents. Cytometry B Clin Cytom 90(1):81–90
9. Paiva B, Vidriales MB, Cervero J, Mateo G, Perez JJ, Montalban MA et al (2008) Multiparameter flow cytometric remission is the most relevant prognostic factor for multiple myeloma patients who undergo autologous stem cell transplantation. Blood 112(10):4017–4023
10. Rawstron AC, Child JA, de Tute RM, Davies FE, Gregory WM, Bell SE et al (2013) Minimal residual disease assessed by multiparameter flow cytometry in multiple myeloma: impact on outcome in the Medical Research Council myeloma IX study. J Clin Oncol 31(20):2540–2547
11. de Tute RM, Rawstron AC, Gregory WM, Child JA, Davies FE, Bell SE et al (2016) Minimal residual disease following autologous stem cell transplant in myeloma: impact on outcome is independent of induction regimen. Haematologica 101(2):e69–e71
12. Kumar S, Paiva B, Anderson KC, Durie B, Landgren O, Moreau P et al (2016) International myeloma working group consensus criteria for response and minimal residual disease assessment in multiple myeloma. Lancet Oncol 17(8):e328–ee46
13. Munshi NC, Avet-Loiseau H, Rawstron AC, Owen RG, Child JA, Thakurta A et al (2017) Association of Minimal Residual Disease with Superior Survival Outcomes in patients with multiple myeloma: a meta-analysis. JAMA Oncol 3(1):28–35
14. Chakraborty R, Muchtar E, Kumar SK, Jevremovic D, Buadi FK, Dingli D et al (2017) Impact of post-transplant response and minimal residual disease on survival in myeloma with high-risk cytogenetics. Biol Blood Marrow Transplant J Am Soc Blood Marrow Transplant 23(4):598–605
15. Royston DJ, Gao Q, Nguyen N, Maslak P, Dogan A, Roshal M (2016) Single-tube 10-Fluorochrome analysis for efficient flow cytometric evaluation of minimal residual disease in plasma cell myeloma. Am J Clin Pathol 146(1):41–49
16. Roshal M, Flores-Montero JA, Gao Q, Koeber M, Wardrope J, Durie BGM et al (2017) MRD detection in multiple myeloma: comparison between MSKCC 10-color single-tube and EuroFlow 8-color 2-tube methods. Blood Adv 1(12):728–732
17. Frassanito MA, Rao L, Moschetta M, Ria R, Di Marzo L, De Luisi A et al (2014) Bone marrow fibroblasts parallel multiple myeloma progression in patients and mice: in vitro and in vivo studies. Leukemia 28(4):904–916
18. Cogbill CH, Spears MD, Vantuinen P, Harrington AM, Olteanu H, Kroft SH (2015) Morphologic and cytogenetic variables affect the flow cytometric recovery of plasma cell myeloma cells in bone marrow aspirates. Int J Lab Hematol 37(6):797–808
19. Joshi R, Horncastle D, Elderfield K, Lampert I, Rahemtulla A, Naresh KN (2008) Bone marrow trephine combined with immunohistochemistry is superior to bone marrow aspirate in follow-up of myeloma patients. J Clin Pathol 61(2):213–216
20. Biran N, Ely S, Chari A (2014) Controversies in the assessment of minimal residual disease in multiple myeloma: clinical significance of minimal residual disease negativity using highly sensitive techniques. Curr Hematol Malig Rep 9(4):368–378
21. Puig N, Sarasquete ME, Balanzategui A, Martinez J, Paiva B, Garcia H et al (2014) Critical evaluation of ASO RQ-PCR for minimal residual disease evaluation in multiple myeloma. A comparative analysis with flow cytometry. Leukemia 28(2):391–397
22. Bai Y, Wong KY, Fung TK, Chim CS (2016) High applicability of ASO-RQPCR for detection of minimal residual disease in multiple myeloma by entirely patient-specific primers/probes. J Hematol Oncol 9(1):107
23. Avet-Loiseau H, Corre J, Lauwers-Cances V, Chretien M-L, Robillard N, Leleu X et al (2015) Evaluation of Minimal Residual Disease (MRD) by Next Generation Sequencing (NGS) is highly predictive of progression free survival in the IFM/DFCI 2009 trial. Blood 126(23):191
24. Ladetto M, Bruggemann M, Monitillo L, Ferrero S, Pepin F, Drandi D et al (2014) Next-generation sequencing and real-time quantitative PCR for minimal residual disease detection in B-cell disorders. Leukemia 28(6):1299–1307
25. Rasche L, Chavan SS, Stephens OW, Patel PH, Tytarenko R, Ashby C et al (2017) Spatial genomic heterogeneity in multiple myeloma revealed by multi-region sequencing. Nat Commun 8(1):268
26. Moreau P, Attal M, Caillot D, Macro M, Karlin L, Garderet L et al (2017) Prospective evaluation of magnetic resonance imaging and [(18)F] Fluorodeoxyglucose positron emission tomography-computed tomography at diagnosis and before maintenance therapy in symptomatic patients with multiple myeloma included in the IFM/DFCI 2009 trial: results of the IMAJEM study. Journal of clinical oncology: official journal of the American Society of Clinical Oncology 35(25):2911–2918

27. Zamagni E, Patriarca F, Nanni C, Zannetti B, Englaro E, Pezzi A et al (2011) Prognostic relevance of 18-F FDG PET/CT in newly diagnosed multiple myeloma patients treated with up-front autologous transplantation. Blood 118(23):5989–5995
28. Bartel TB, Haessler J, Brown TL, Shaughnessy JD Jr, van Rhee F, Anaissie E et al (2009) F18-fluorodeoxyglucose positron emission tomography in the context of other imaging techniques and prognostic factors in multiple myeloma. Blood 114(10):2068–2076
29. Usmani SZ, Mitchell A, Waheed S, Crowley J, Hoering A, Petty N et al (2013) Prognostic implications of serial 18-fluoro-deoxyglucose emission tomography in multiple myeloma treated with total therapy 3. Blood 121(10):1819–1823
30. Davies FE, Rosenthal A, Rasche L, Petty NM, JE MD, Ntambi JA et al (2018) Treatment to suppression of focal lesions on positron emission tomography-computed tomography is a therapeutic goal in newly diagnosed multiple myeloma. Haematologica 103(6):1047–1053
31. Sachpekidis C, Hillengass J, Goldschmidt H, Mosebach J, Pan L, Schlemmer HP et al (2015) Comparison of (18)F-FDG PET/CT and PET/MRI in patients with multiple myeloma. Am J Nucl Med Mol Imaging 5(5):469–478
32. San MJ (2014) Multiple myeloma: a model for scientific and clinical progress. Hematology Am Soc Hematol Educ Program 2014(1):1–7
33. Paiva B, Corchete LA, Vidriales MB, Puig N, Maiso P, Rodriguez I et al (2016) Phenotypic and genomic analysis of multiple myeloma minimal residual disease tumor cells: a new model to understand chemoresistance. Blood 127(15):1896–1906
34. Keats JJ, Chesi M, Egan JB, Garbitt VM, Palmer SE, Braggio E et al (2012) Clonal competition with alternating dominance in multiple myeloma. Blood 120(5):1067–1076
35. Ghosh A, Carreau N, Moscatello A, Rahman A, Qi J, Kim-Schulze S et al (2015) Flow cytometry based detection of MRD in bone marrow of patients with multiple myeloma: a comparison between fluorescent-based cytometry versus cytof. Blood 126(23):4195
36. Mithraprabhu S, Khong T, Ramachandran M, Chow A, Klarica D, Mai L et al (2017) Circulating tumour DNA analysis demonstrates spatial mutational heterogeneity that coincides with disease relapse in myeloma. Leukemia 31(8):1695–1705
37. Mills JR, Barnidge DR, Dispenzieri A, Murray DL (2017) High sensitivity blood-based M-protein detection in sCR patients with multiple myeloma. Blood cancer journal. 7(8):e590
38. Martinez-Lopez J, Lahuerta JJ, Pepin F, Gonzalez M, Barrio S, Ayala R et al (2014) Prognostic value of deep sequencing method for minimal residual disease detection in multiple myeloma. Blood 123(20):3073–3079
39. Silvennoinen R, Lundan T, Kairisto V, Pelliniemi TT, Putkonen M, Anttila P et al (2014) Comparative analysis of minimal residual disease detection by multiparameter flow cytometry and enhanced ASO RQ-PCR in multiple myeloma. Blood Cancer J 4:e250
40. Lohr JG, Kim S, Gould J, Knoechel B, Drier Y, Cotton MJ et al (2016) Genetic interrogation of circulating multiple myeloma cells at single-cell resolution. Science Translat Med 8(363):363ra147
41. Ortega MA, Poirion O, Zhu X, Huang S, Wolfgruber TK, Sebra R et al (2017) Using single-cell multiple omics approaches to resolve tumor heterogeneity. Clin Transl Med 6(1):46
42. Palumbo A, Chanan-Khan A, Weisel K, Nooka AK, Masszi T, Beksac M et al (2016) Daratumumab, Bortezomib, and dexamethasone for multiple myeloma. N Engl J Med 375(8):754–766
43. Dimopoulos MA, Oriol A, Nahi H, San-Miguel J, Bahlis NJ, Usmani SZ et al (2016) Daratumumab, Lenalidomide, and dexamethasone for multiple myeloma. N Engl J Med 375(14):1319–1331
44. Avet-Loiseau H, Casneuf T, Chiu C, Laubach JP, Lee J-J, Moreau P et al (2016) Evaluation of minimal residual disease (MRD) in relapsed/refractory multiple myeloma (RRMM) patients treated with Daratumumab in combination with Lenalidomide plus dexamethasone or Bortezomib plus dexamethasone. Blood 128(22):246
45. Paiva B, Martinez-Lopez J, Vidriales MB, Mateos MV, Montalban MA, Fernandez-Redondo E et al (2011) Comparison of immunofixation, serum free light chain, and immunophenotyping for response evaluation and prognostication in multiple myeloma. Journal of clinical oncology : official journal of the American Society of Clinical Oncology 29(12):1627–1633
46. Korthals M, Sehnke N, Kronenwett R, Bruns I, Mau J, Zohren F et al (2012) The level of minimal residual disease in the bone marrow of patients with multiple myeloma before high-dose therapy and autologous blood stem cell transplantation is an independent predictive parameter. Biology of blood and marrow transplantation : journal of the American Society for Blood and Marrow Transplantation 18(3):423–31.e3
47. Paiva B, Gutierrez NC, Rosinol L, Vidriales MB, Montalban MA, Martinez-Lopez J et al (2012) High-risk cytogenetics and persistent minimal residual disease by multiparameter flow cytometry predict unsustained complete response after autologous stem cell transplantation in multiple myeloma. Blood 119(3):687–691
48. Roussel M, Lauwers-Cances V, Robillard N, Hulin C, Leleu X, Benboubker L et al (2014) Front-line transplantation program with lenalidomide, bortezomib, and dexamethasone combination as induction and consolidation followed by lenalidomide maintenance in patients with multiple myeloma: a phase II study by the Intergroupe francophone du Myelome. Journal of clinical oncology : official journal of the American Society of Clinical Oncology. 32(25):2712–2717

49. Ferrero S, Ladetto M, Drandi D, Cavallo F, Genuardi E, Urbano M et al (2015) Long-term results of the GIMEMA VEL-03-096 trial in MM patients receiving VTD consolidation after ASCT: MRD kinetics' impact on survival. Leukemia 29(3):689–695
50. Korde N, Roschewski M, Zingone A, Kwok M, Manasanch EE, Bhutani M et al (2015) Treatment with carfilzomib-Lenalidomide-dexamethasone with Lenalidomide extension in patients with smoldering or newly diagnosed multiple myeloma. JAMA Oncol 1(6):746–754
51. Ludwig H, Greil R, Masszi T, Spicka I, Shpilberg O, Hajek R et al (2015) Bortezomib, thalidomide and dexamethasone, with or without cyclophosphamide, for patients with previously untreated multiple myeloma: 5-year follow-up. Br J Haematol 171(3):344–354
52. Oliva S, Gambella M, Gilestro M, Muccio VE, Gay F, Drandi D et al (2017) Minimal residual disease after transplantation or lenalidomide-based consolidation in myeloma patients: a prospective analysis. Oncotarget 8(4):5924–5935
53. Paiva B, Cedena MT, Puig N, Arana P, Vidriales MB, Cordon L et al (2016) Minimal residual disease monitoring and immune profiling in multiple myeloma in elderly patients. Blood 127(25):3165–3174
54. Attal M, Lauwers-Cances V, Hulin C, Leleu X, Caillot D, Escoffre M et al (2017) Lenalidomide, Bortezomib, and dexamethasone with transplantation for myeloma. N Engl J Med 376(14):1311–1320

Index

J. A. Aguirre-Ghiso (ed.), *Biological Mechanisms of Minimal Residual Disease and Systemic Cancer*, Advances in Experimental Medicine and Biology 1100,
https://doi.org/10.1007/978-3-319-97746-1

The manufacturer's authorised representative in the EU is Springer Nature Customer Service Centre GmbH, Europaplatz 3, 69115 Heidelberg, Germany. If you have any concerns regarding our products, please contact ProductSafety@springernature.com

Printed and bound by CPI Group (UK) Ltd, Croydon, CR0 4YY
30/09/2026
02225471-0019